Digital Signal
Processing Design

Andrew Bateman PhD
University of Bristol

Warren Yates PhD
University of Technology
Sydney

1996

Pitman

Pitman Publishing
128 Long Acre London WC2E 9AN

© A. Bateman & W. Yates 1988

First published in Great Britain 1988

British Library Cataloguing in Publication Data

Bateman, Andrew
   Digital signal processing design.
   1. Digital signals. Processing. Applications
   I. Title   II. Yates, Warren
   621.38'043

   ISBN 0-273-02787-5

ISBN 0 273 02787 5

Printed and bound in Great Britain at The Bath Press, Avon

Contents

We would like to acknowledge and thank James Marvill for the considerable time and effort he gave to the preparation and evaluation of the example programs to be found in the text.

To both our families, especially Jacqui and Judy

Introduction

In 1979, Intel introduced the first microprocessor with an architecture and instruction set specifically tailored for Digital Signal Processing (DSP) applications. Since then, VLSI DSP chips have been launched by Texas Instruments, IBM, Analog Devices, Motorola, NEC, STL, AT&T, Inmos, and many others. The rapid growth in the exploitation of DSP technology is not surprising considering the commercial advantages in terms of fast, flexible, precise and potentially low-cost design capabilities offered by these devices, with many companies already exploiting DSP technology in the areas of communications, control, instrumentation, and more recently, talking dolls!

This book is aimed primarily at the engineer or designer who, although familiar with the theory and practice of analog system design, has had little or no exposure to DSP technology and wishes to acquire skill in this area with the minimum possible investment of time and effort. It is also intended as a general handbook of processing algorithms and circuit design techniques for the experienced engineer, forming the basis for more sophisticated system development in application areas falling outside the scope of this text.

The guiding principle in writing the book has been the belief that acquiring competence in any new field requires, above all, the building of confidence, and that confidence in turn is best developed by successful implementation in hardware of one's own designs at an early stage in the learning process. With this in mind the material is presented in the form of specific algorithms and explanatory material on hardware implementation so that the reader, with access to appropriate development systems, can tackle a section of the book and immediately try out a related design. Most of the material is drawn from the authors' recent experience of grappling with concepts and applications of digital signal processing from both an undergraduate teaching and postgraduate research standpoint.

Considerable attention has been paid to the organisation of the material so as to ensure a progressive development of understanding of the theoretical background to DSP, with sufficient theory being presented to allow the reader to modify, extend and invent algorithms without running foul of fundamental theoretical constraints.

The book falls naturally into three sections. The first provides the context for the remainder, outlining the fundamental differences in approach between analog and digital signal processing design which result from the discrete rather than continuous processing of signal samples. This is followed by a brief description of the architecture, instruction sets and performance of many typical DSP chips. The middle section, which constitutes the bulk of the book, covers a number of general application areas, providing, in effect, a library of DSP algorithms accompanied in many cases with implementation examples based upon the Texas Instruments TMS320 series of DSP devices. The final section is devoted to hardware design. A number of practical interface circuits are presented for the TMS320 series which provide a rapid and economical path for the design engineer into hardware implementation and evaluation of DSP for his or her given application. A more detailed summary of the contents of individual chapters is given below.

Chapter Contents

Chapter 1 begins by outlining the basic operation of DSP and discusses the advantages and limitations of DSP system implementation in comparison to traditional analog techniques. A typical DSP chip is described both in terms of its architecture and instruction set, and the differences between DSP chips and general-purpose microprocessors are highlighted. The various phases of a DSP design from specification to implementation are described along with typical development aids. Some basic rules are developed for assessing the feasibility of using a DSP chip with a given performance for a particular processing application.

Chapter 2 is a guided tour through continuous and discrete signal theory. The consequences of waveform sampling in terms of reconstruction and aliasing are explained, and the theoretical justification for DSP operations such as interpolation and decimation is presented. A brief section on the use of the z-transform as a means of representing discrete time signals and processes is included as a basis for the understanding of filtering and the fast Fourier transform techniques considered in subsequent chapters.

It is good practice to synthesize complex software systems from simple modules that have been individually tested. *Chapter 3* proposes a 'toolbox' of basic DSP modules from which larger systems can be constructed. Modules for input/output, signal delay, rectification, quantization, multiplication, division, exponentiation, and root finding, etc, are presented along with utility modules such as counters and timers. In each case, example programs for the popular TMS320 family of processors are given.

Chapter 4 is devoted to the topic of filtering. Filters can be implemented as recursive structures (those employing feedback), or non-recursive structures (those without feedback), and can be based on existing analog designs or be developed as discrete time filters from the outset. This chapter guides the engineer through the process of matching filter type to system requirements emphasising the trade-offs between performance, processing time and memory requirements. Reference is made to the vast wealth of software filter design packages available commercially or to be found in the open literature.

Chapter 5 deals with the concept of spectral analysis and the relationship between the continuous and discrete versions of the Fourier transform. This is followed by a review and examples of fast Fourier transform algorithms and techniques. The remainder of the chapter is concerned with the application of spectral analysis techniques in areas such as speech recognition, autocorrelation and system identification.

In *Chapter 6*, the focus is on modules for the synthesis of communications systems. Algorithms and code examples are given for waveform generators, envelope and phase detectors, amplitude frequency and phase modulators, specialised filters, etc., together with a number of practical examples showing the interconnection of modules to realize the more complex communication systems.

The final chapter, *Chapter 7*, deals with the hardware aspects of DSP design starting with an overview of DSP peripheral devices such as A/D converters, memory blocks, etc. This is followed by a section describing a series of low-cost prototyping systems developed primarily for the TMS320 processor range including a PROM emulation board and general-purpose analog interface board. The chapter concludes with a section on device support tools such as simulators and emulators, and provides some insight into the future trends in the DSP market place.

Computer Systems series

Consultant Editor: John Freer, Principal Consultant, Software
Sciences Limited

1 Fundamentals

1.1 Aims of the book

This book is intended as a self-study course in Digital Signal Processing (DSP) for the analog designer. It forms a complete introduction to the subject in the sense that theory, software design, and hardware design are given a balanced coverage. The book is oriented towards communications applications that are to be implemented on one or more of the general-purpose monolithic DSP devices introduced into the market over the last few years. Examples are the *Texas Instruments TMS320 family, the Analog Devices ADSP2100, NEC's MPD7720 family, Fujitsu's MB8764, Motorola's DSP56000, AT&T's WE DSP16 and WE DSP32, and Thomson's 68931.* In this introductory chapter we make few assumptions about prior knowledge but move at a fairly rapid pace through a body of prerequisite material explaining terminology and concepts that will be needed in Chapters 3 and beyond. The bulk of the book is devoted to the description of **algorithms** for particular processing tasks. It is hoped that this collection of techniques culled from each author's own experience and the considerable literature on the subject will enable the reader to proceed quickly and painlessly to successful implementation.

In designing with DSP devices it is relatively straightforward to get a process working once the algorithm is known. The bulk of the development cycle is spent in making refinements to improve speed and performance. This book is primarily aimed at the algorithm selection and initial implementation phase. Knowing how to make further improvements comes with an intimate knowledge of the particular device being used.

Although most of the example programs are for the Texas Instruments TMS32010 and TMS32020 processors, the algorithms themselves are described in general terms and it is recommended that the book be read in conjunction with the user guide of the DSP device to be used.

1.2 What is digital signal processing?

For the majority of applications, DSP involves the replacement of analog elements such as amplifiers, modulators, and filters by the sub-system shown in Fig 1.1. The waveform to be processed is filtered, then sampled, then converted to a digital representation, and finally input to the signal processor. *All subsequent wave form manipulation is then implemented in software.* The processed signal samples in the form of digital words are finally converted back to analog samples and passed through a reconstruction filter to recover the signal as a continuous waveform.

Fig 1.1 A digital signal processing system

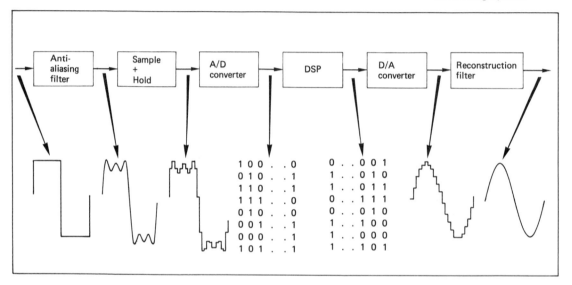

It is important to distinguish DSP from data processing. Data processing involves manipulating information which is already digitized and which may or may not represent a current action and for which the need for a result is not a function of real time. DSP has to do with real-time analog signals which just happen to be processed digitally. While conventional microprocessors can be used for both signal and data processing, their architectures and instruction sets limit their usefulness as signal processors to applications where the signals have bandwidths of a few hundred hertz.

Speech analysis was the driving force behind initial attempts to process signals digitally. The exacting tolerances demanded of filters in speech processing systems simply could not be maintained over time with analog techniques, subject as they are to temperature drift, component tolerances and ageing. Digital processing operations, in contrast, consist of nothing more than sequences of binary multiplication, addition, subtraction, etc, making the outcome entirely predictable and thus reproducible. In the sixties, DSP hardware used discrete components and consequently, because of the high cost and volume, its application could only be justified for very specialised requirements (or large-budget research programmes!). In the seventies, monolithic components for some of the DSP subsystems appeared, primarily dedicated digital multipliers and address generators (cf. section 1.6.4.) and DSP systems could be implemented using bit slice microprocessors. The breakthrough for mass exploitation of DSP techniques came in 1979 when Intel introduced the 2920, a completely self-contained signal processing device in a 40-pin DIP package incorporating on-board program EPROM, data RAM, A/D and D/A converters, and an architecture and instruction set powerful enough to implement a full duplex 1200 bps modem, including transmit and receive filters.

Since the Intel 2920 there have been two further generations of general-purpose signal processing devices, with a fourth generation announced. All the major device manufacturers have products either already released or under development, reflecting their confidence in an enormous growth of demand in the late eighties and early nineties, paralleling the spectacular success of microprocessors in the seventies.

1.2.1 Advantages of digital signal processing

What are the advantages of DSP over conventional analog design? The first, that of precision and repeatability, has already been mentioned. Some of the others are listed in Table 1.1. In common with all microprocessor-based systems, the design can be simulated in software and easily modified, thus greatly reducing product development time. A second major advantage of DSP-device-based signal processing is the property that the complexity of the processing becomes independent of the hardware. Many operations that are difficult to achieve with analog processing, such as linear phase filters and precision non-linear function generation (discussed in Chapters 3 and 4), are straightforward using digital techniques. It may be some time before equipment designers fully exploit this capability, but when this happens we can expect to see some radically new approaches to solving old problems.

Some of the many applications areas are given in Table 1.2. A likely trend in the future will be the appearance of DSP products tailormade for particular applications as opposed to the general-purpose families currently available. One such product range is the ZORAN series with its different devices for frequency domain and time domain processing.

Advantages of Digital Signal Processing
Precision, hence predictability.
Stability (no drifting with temperature or age).
Controlled accumulation of noise.
Hardware cost independent of complexity.
Easily modified.
Can be simulated.
Easy to debug, hence short development time.
Simpler hardware (lower cost and greater reliability).
Functions implementable that are analog-impractical.

Limitations of Digital Signal Processing
Bandwidth limited by sampling rate.
Limited dynamic range.
Quantization noise.
Round-off errors.

Table 1.1 Advantages and limitations of DSP

1.2.2 Inherent limitations of digital signal processing

Many of the desirable properties of DSP stem from the control they give the designer over performance. In analog design, performance deterioration with time is only controllable on a statistical basis. With DSP the designer builds in a certain degradation at the outset, but is confident that it will never get worse. The initial degradation is a consequence of the loss of information that is inherent in sampling and quantizing real signals. **Quantization** is the process of representing an analog sample by the *nearest* level that exactly corresponds to an integer scale (Fig 1.2). This necessarily introduces quantization noise, defined as the difference between the true value and the actual

Fig 1.2 The error introduced by quantization

Fig 1.3 Input, output and error waveforms for the quantization characteristic in Fig 1.2

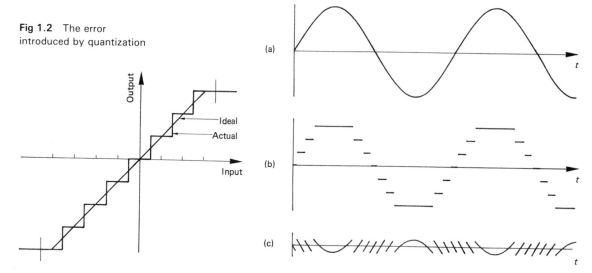

General-purpose DSP	Graphics/Imaging	Instrumentation
Digital Filtering	3-D Rotation	Spectrum Analysis
Convolution	Robot Vision	Function Generation
Correlation	Image Transmission/	Pattern Matching
Hilbert Transforms	Compression	Seismic Processing
Fast Fourier Transforms	Pattern Recognition	Transient Analysis
Adaptive Filtering	Image Enhancement	Digital Filtering
Windowing	Homomorphic Processing	Phase-Locked Loops
Waveform Generation	Workstations	
	Animation/Digital Map	
Voice/Speech	**Control**	**Military**
Voice Mail	Disk Control	Secure Communications
Speech Vocoding	Servo Control	Radar Processing
Speech Recognition	Robot Control	Sonar Processing
Speaker Verification	Laser Printer Control	Image Processing
Speech Enhancement	Engine Control	Navigation
Speech Synthesis	Motor Control	Missile Guidance
Text to Speech		Radio Frequency Modems
Telecommunications		**Automotive**
Echo Cancellation	FAX	Engine Control
ADPCM Transcoders	Cellular Telephones	Vibration Analysis
Digital PBXs	Speaker Phones	Antiskid Brakes
Line Repeaters	Digital Speech	Adaptive Ride Control
Channel Multiplexing	Interpolation (DSI)	Global Positioning
1200 to 19200 bps Modems	X.25 Packet Switching	Navigation
Adaptive Equalizers	Video Conferencing	Voice Commands
DTMF Encoding/Decoding	Spread Spectrum	Digital Radio
Data Encryption	Communications	Cellular Telephones
Consumer	**Industrial**	**Medical**
Radar Detectors	Robotics	Hearing Aids
Power Tools	Numeric Control	Patient Monitoring
Digital Audio/TV	Security Access	Ultrasound Equipment
Music Synthesizer	Power Line Monitors	Diagnostic Tools
Educational Toys		Prosthetics
		Fetal Monitors

Table 1.2   Some typical applications of DSP

value of a waveform that would be observed if the digital representation were immediately applied to the complementary D/A convertor (Fig 1.3). The greater the number of bits used to represent a waveform sample the smaller will be the quantization noise. The signal-to-noise ratio for a sine wave which just fits in the allowable voltage swing of the D/A converter is given by

$$\text{SNR} = 6.02\,n + \log_{10}(1.5)\ \text{dB}$$

where n is the number of bits in the digital representation. Too low a value of n also restricts the dynamic range. The largest *signed number* representable with n bits is 2^{n-1} and the smallest is 1, giving a dynamic range of $20(n-1)\log 2$ or 42 dB for 8 bits, 66 dB for 12 bits, and 90 dB for 16 bits. (For *unsigned number* representation, the dynamic range obtainable is increased by 6 dB, cf. section 1.5.1). Overflow and underflow associated with integer arithmetic operations within the processor are a further constraint.

The speed of processing and thus how fast the digital information can be handled determines the rate at which the analog signal can be sampled. Strictly bandlimited signals can be sampled without information loss if the sampling rate is more than twice the bandwidth. However most real life signals have to be filtered to make the bandlimited assumption even approximately true, and the resulting distortions that follow from the filtering itself and the high-frequency components that pass through the filter must be accepted.

A further potential source of degradation is due to timing jitter on the sampling pulses. This will be negligible if a crystal-controlled clock is used.

The key requirements of the processor are, thus, speed and word length and both adversely affect cost. Currently available DSP devices represent data samples as either 16-bit integers or 32-bit floating-point numbers, and use clock frequencies up to 42 MHz to give 50 to 200 nanosecond cycle times. Depending on the complexity of processing and the number of signals being concurrently processed these devices are usable with signals with equivalent baseband bandwidths of up to a few hundred kHz.

1.3   Digital signal processing hardware

We have already seen from Fig 1.1 that a DSP system requires filters, A/D and D/A converters as well as the actual processor IC. The processor itself also needs a number of supporting ICs and these will be discussed briefly in this section. Chapter 7 tackles the question of hardware design in detail. Fig 1.4 illustrates how a DSP system is configured around the Analog Devices ADSP2100 processor. The major external blocks needed are *program memory*, *data memory*, *clock*, and *interrupt* modules, and a variety of external device interfaces, in this case a 16-bit *parallel port*.

The different DSP ICs vary as to the extent that memory and peripheral devices are provided on-chip. At one extreme the NEC MPD7720 has all of its program and data memory on-chip (including a block of data EPROM for coefficients), thus avoiding the need for an external bus and minimising pin count. At the other extreme the Analog Devices ADSP2100 has no on-chip memory of either type,

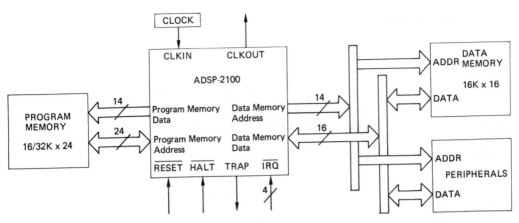

Fig 1.4 External interfaces for the Analog Devices ADSP2100

but brings out address, data and handshake lines to interface with a block of $16 \, \text{K} \times 16$-bit-wide data memory and a block of $16 \, \text{K} \times 24$-bit-wide program memory. The usual configuration is to have some on-chip program (EP)ROM and data RAM with provision for external expansion, possibly with a performance penalty of extra machine cycles for external memory accesses. In the Texas Instruments TMS32010 the program and data buses and address lines are multiplexed for external access to off-chip memory of both types and external peripherals. The TMS32010 specifies external RAM with an access time of less than 100 ns. Other devices make provision for access to low-speed memory. Multiplexing of external data and program memory buses is also used on the Motorola DSP56000 to reduce pin count.

Since the demise of the Intel 2920 no DSP device has included on-board A/D and D/A converters, the justification being the desire on the part of the designers not to lock users into the limitations of a particular A/D and D/A converter. The analog signal interface must therefore be provided by the user. A/D and D/A converters can either connect to the data bus, having a dedicated address (memory mapping), or use a dedicated parallel or serial interface if one is available.

Because constant-rate sampling is an essential feature of signal processing, DSP devices have either an on-board timer or one or more external interrupt request lines, or both. External interrupts, and those generated internally by on-chip peripherals, are prioritized and may be masked out under software control.

All the available devices use a single 5 V supply and are implemented in either NMOS or CMOS technology, with TTL external interfaces and a power consumption of less than a few watts. Package details are summarised in Table 1.3.

1.4 DSP device architectures

Fig 1.5 shows the internal architecture of the Analog Devices ADSP2100. All DSP ICs employ a similar interconnection of functional blocks although the nature and performance of equivalent sub-systems in different devices can vary markedly. Compared with conventional microprocessors the most striking feature is the use of parallelism and pipelining to improve speed. Most DSP devices use the **Harvard architecture**, characterised by the separation of *program memory* from *data memory*. This entails separate *address* and *data buses* for both program and data memory, which means that data fetches can be concurrent with the fetching of the next program instruction. DSP devices have a number of dedicated processing units all of which can operate independently and (with some restrictions) concurrently. Always present in some form is an *arithmetic logic unit* (ALU) for performing addition, subtraction and logical operations, a *shifter* for scaling data on its way to or from other processing units, a *multiplier* (MAC), and at least one ALU dedicated to *address generation* tasks.

In the Analog Devices ADSP2100 each of the processor units (the ALU, MAC and shifter) has its own *input and output registers* which may be written to or read (for the next operation or from the

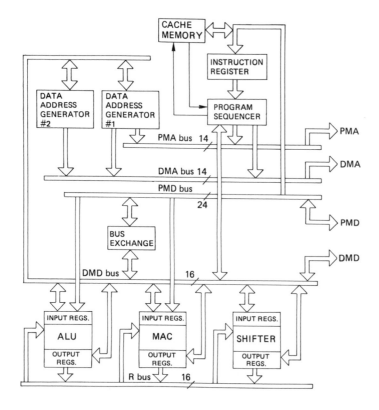

Fig 1.5 ADSP2100 internal architecture

Processor	Pin count	Package
Texas Instruments TMS32010	40	DIP
Texas Instruments TMS32020	68	PGA
Texas Instruments TMS320C25	68	PLCC
NEC MPD7720	28	DIP
NEC MPD77230	64	PGA
Fujitsu MB8764	88	PGA
Analog Devices ADSP2100	100	PGA
Motorola DSP56000	88	PGA
Thomson 68931	84	PGA
AT&T WE DSP16	84	PLCC
AT&T WE DSP32	40(100)	DIP(PGA)

Table 1.3 DSP IC packaging details

previous operation respectively) at the same time as they are used for input or output in the current operation. This form of **pipelining** is illustrated in Fig 1.6. As a result a 2100 instruction can provide for any of the combinations of operations listed in Table 1.4 to be executed in a single cycle.

All the DSP devices employ equivalent strategies for pipelining, although not necessarily by means of individual input and output registers for each processing sub-system, or such unrestricted register-to-register transfer options.

As in conventional microprocessors the processing sub-systems interwork under the overall control of the **program sequencer,** which has provision for condition testing, context saves following an interrupt, handshaking with external devices, and sharing resources with external processors on a master or slave basis.

Because signal processing algorithms tend to involve repeated

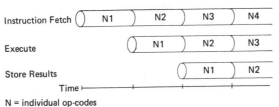

Fig 1.6 Three-stage pipeline processing: several tasks are performed simultaneously in each instruction cycle

- any ALU,MAC or shifter operation.
- any register-to-register move.
- any data or program memory read or write.
- a computation with a register-to-register move.
- a computation with any memory read or write.
- a computation with a read from both of two external memories.

Table 1.4 Single-cycle operations supported by the Analog Devices ADSP2100 instruction set

cycles of multiplication and accumulation of data elements with coefficients stored in different parts of memory, the efficiency with which a data memory address can be generated has a major bearing on processor speed. To minimise processing time for a Fast Fourier Transform for example, it is useful to be able to sequentially generate addresses in a circular buffer for signal input and output, to sequentially select every nth sample for decimation, and to reorder data by generating bit-reversed addresses (see Chapter 5). In the more advanced DSP devices all these features are provided by the address generator sub-system.

Before going into further detail on architecture it is necessary to digress for a moment to consider how signal samples might be internally represented and manipulated within a digital signal processing system. We will then be in a position to appreciate some of the reasons for the choices made by device designers. The discussion of hardware resumes in the next section but one.

1.5   Internal representation of signal samples and coefficients

1.5.1   Conventions in use

When digitizing waveform samples the obvious convention for mapping the sample values to numbers, as is required in the A/D convertor, is to use the rule that a numerical value of N represents any signal sample lying in the range $(N - 1/2)D$ to $(N + 1/2)D$ volts where D is the step size voltage. In the D/A converter the digital value N causes an output level of ND volts. Table 1.5 demonstrates this format for a 3-bit converter with a step size of $1/8$, giving an input range for less than $1/16$ quantization error of zero to $15/16$ volts. For input levels which are integer multiples of $1/8$ the quantization error is zero.

This format for representing a waveform sample is perfectly satisfactory for positive waveform samples, but makes no provision for negative input values. One way of handling signed numbers is to simply designate the most significant bit (MSB) as the sign bit (0 for positive and 1 for negative), with the magnitude being conveyed by the remaining bits using the above unsigned convention. This scheme, which is called **sign magnitude**, has the disadvantage of having two representations of zero.

A more useful format which is used by most A/D converter ICs is **offset binary**, where the highest positive sample is assigned the highest binary number (all 1s) with each quantization step reduction in the signal being reflected in a unit decrement in the digital output.

Input range (volts)	Numerical representation	Binary equivalent	Output (volts)
13/16 to 15/16	7	111	7/8
11/16 to 13/16	6	110	6/8
9/16 to 11/16	5	101	5/8
7/16 to 9/16	4	100	4/8
5/16 to 7/16	3	011	3/8
3/16 to 5/16	2	010	2/8
1/16 to 3/16	1	001	1/8
0 to 1/16	0	000	0

Table 1.5 Conversion rule for 3-bit unsigned A/D and D/A converters with a step size of 1/8

Input level	Sign Magnitude	Offset binary	2's complement
7/8 to 9/8	–	–	–
5/8 to 7/8	011	111	011
3/8 to 5/8	010	110	010
1/8 to 3/8	001	101	001
– 1/8 to 1/8	000 or 100	100	000
– 3/8 to – 1/8	101	011	111
– 5/8 to – 3/8	110	010	110
– 7/8 to – 5/8	111	001	101
– 9/8 to – 7/8	–	000	100

Table 1.6 Three different conventions for representing signed signal samples

This results in the all-zero code being associated with a negative sample one quantization step larger in magnitude than the peak positive value. A comparison between sign magnitude and offset binary conventions for a 3-bit converter with a step size of 1/4 is given in Table 1.6. Note that in offset binary the MSB still indicates polarity and that there is asymmetry in the voltage level which can be represented.

Before the offset binary numbers produced by the A/D converter can be used for processing they must be converted into a form which can support arithmetic operations. We would like to be able to add the numerical representations of signal samples so as to get the equivalent of adding the signal samples. This is not the case with offset binary. Referring to Table 1.6, signal samples of − 1/4 and − 1/2 have representations of 011 and 010 respectively which, when added, give 101 which is equivalent to 1/4 instead of − 3/4 as we would wish.

The solution is to represent positive samples using the unsigned convention and negative samples by the number which would have to be added to the magnitude representation to give zero. For example if + V is 0110 then − V is 1010, since

$$0110 + 1010 = (1)0000$$

To obtain the binary code for a negative sample the equivalent positive sample representation is complemented (this is the so-called 1's complement) and the result incremented by 1 to form the **2's complement**. In our example we have

Magnitude	1's complement	2's complement
0110	1001	1010

In Table 1.6 the 2's complement representation can be compared with offset binary. Notice that the conversion from offset binary to 2's complement and the reverse involves only a reversal of the MSB.

To summarise, signal samples are represented in DSP memory by binary numbers in 2's complement format. The MSB indicates the polarity, 0 for positive and 1 for negative.

Not all data values that need to be represented in signal processing are signed. An example is loop counts. In what follows we consider in more detail how addition/subtraction and multiplication are performed, first by using unsigned format, then the 2's complement format, for signed numbers.

1.5.2 Addition and subtraction of unsigned operands

An n-bit adder with *carry* and provision for *carry-in* can be used for adding numbers in unsigned format with widths which are integer multiples of n. Consider the use of a 4-bit adder to add the 8-bit numbers 00110110 and 01011101. In the first iteration the two low-order nibbles (4 bits) are loaded into the adder giving

```
    0    carry-in
 0110    x
 1101    y
 ─────
10011
─────
c3210    bit position
```

In the next iteration the higher-order nibbles are added with the previous carry, giving

```
    1    carry-in
 0011    x
 0101    y
 ─────
01001
─────
c3210    bit position
```

Not all DSP devices make provision for multiple precision arithmetic by providing the necessary carry register and allowing for carry-in.

All however set a flag when an *overflow* or an *underflow* occurs, that is when the result of an addition or subtraction of the two n-bit numbers requires more than n bits.

Some processors use a double-length accumulator (i.e. twice the length of the data word) for addition/subtraction. Single precision numbers can then be loaded into either the low or the high half of the accumulator. If addition is performed with two numbers in the low half of the accumulator, the carry bit will be the LSB of the high half. If the numbers are initially loaded into the high half and addition is performed, a carry is signalled by the setting of the overflow flag. Sometimes a separate overflow flag is provided for a low-half to high-half overflow.

1.5.3 Addition and subtraction in 2's complement format

An example of a straightforward 2's complement addition has already been given. Consider now what happens when a carry is generated in 2's complement addition. A carry in this case does not necessarily imply an overflow. Consider the following 4-bit (3 + sign) addition:

$$
\begin{array}{ll}
1110 & (-2) \\
1110 & + (-2) \\
\hline
11100 & = (-4) \\
\hline
c3210 & \textit{bit position}
\end{array}
$$

This gives the correct result and the carry is simply an extra sign bit which can be ignored. The next example:

$$
\begin{array}{ll}
1000 & (-8) \\
1110 & + (-4) \\
\hline
10110 & = (+6) + \text{carry bit} \\
\hline
c3210 & \textit{bit position}
\end{array}
$$

gives a carry bit which *does* signify an overflow. Overflow for 2's complement is signified by the carry bit being different from the MSB. Because an overflow in 2's complement format has the rather drastic effect of changing a large positive number into a large negative one, or vice versa, DSP devices have a facility for operating in **saturation mode**. It means that when an overflow occurs the register contents are replaced by the largest number of the appropriate sign that can be represented. The operation is similar to an analog limiter (cf. overflow control, Chapter 3). In the above example the result register would contain 1000 (-8) if saturation mode had been enabled.

If 2's complement addition is to be performed using a double-length accumulator with the operands loaded into the lower half, it is necessary to extend the sign bit to the left. For example, the 12-bit representation of -5 is *111111111011* compared with the 4-bit representation 1011. In order to be able to handle both signed and unsigned addition/subtraction, provision is made in the processor instruction set for specifying whether sign extension is to apply when an n-bit operand is loaded into a register of length greater than n.

Double precision signed numbers can be added using carry and carry-in bits as illustrated above for the unsigned case, provided that sign extension over the double width is applied.

1.5.4 Multiplication of signed and unsigned operands

An unsigned binary number can be multiplied by $2^{\pm n}$ by an n-bit left/right shift. Multiplication by an integer can thus be implemented as a sequence of shifts and adds. To multiply a number by 6, for example, it must be left-shifted 2 bits ($\times 4$) and added to the original number left-shifted 1 bit ($\times 2$). To multiply by 0.75 requires 2 right shifts added to 1 right shift.

All DSP devices are equipped with a hardware multiplier which performs all the necessary shifts and adds in a single machine cycle. The multiplier accepts two n-bit operands and produces a $2n$-bit result. If the two inputs were 0100 and 1100, the result would be formed by the following addition:

```
    010000
   0100000
```
————————
```
  00110000
```
————————
76543210   bit position

Note that there can be no overflow if the result register is $2n$ bits wide.

Some care must be taken when performing a multiplication to keep track of the **implied binary point**. If the numbers 0100 and 1100 represent 4 and 12 respectively, the binary point is immediately to the right of the LSB. Following multiplication (result: $00110000 = 48$) the position of the binary point is unchanged.

Q notation   The Q notation is commonly adopted for specifying the position of the implied binary point, such that a binary number in Qn format is defined as having n bits to the right of the binary point. Thus all the numbers in the above example are in Q0 format.

The general rule for multiplication is that if the multiplicands are in Qn and Qm format respectively the result will be in $Q(n + m)$

format. Thus if one of the multiplicands is in Q0 format (i.e. an integer) then the result will have the format of the other multiplicand. Suppose 01.110 represents 1.75 (i.e. the format is Q3). Multiplying by 2, represented in Q0 format as 0010, gives 011.100. The result has the same format as the non-integer operand, i.e. Q3, so the result is correctly interpreted as 3.5.

A useful convention for **fixed-point** digital signal processing is to interpret signal samples as integers (we have already seen how the integer can be interpreted in terms of the number of quantization steps), hence in Q0 format, and to represent coefficients in sub-unitary format, that is to allow only coefficient values less than unity. All coefficients can then be represented in binary format with the implied binary point to the left of the MSB. A 16-bit-wide data element using this format could thus represent an unsigned coefficient from 0 to $(1 - 2^{-16})$ using Q16 format or signed coefficient from -1 to $(1 - 2^{-15})$ using Q15 format.

If this convention is adopted, then multiplication of a *signed* signal sample by a *signed* coefficient will yield a result in the double width accumulator with binary point between the high and low halves. The 32-bit result in the double-length accumulator will be in Q15 format. A *unit left shift* followed by storing from the *high* half of the accumulator will give the 16-bit result in Q0 format. Many DSP devices have provision for a software selectable automatic left shift prior to storage from the high half of the accumulator. [Note: it is equally valid to interpret the signal samples as Q15 representations of voltages normalized to the peak magnitude output of the D/A converter. A multiplication now gives a result in Q30 format (the sign bit is duplicated) and again a one-bit left shift before storage from the high half gives the result in Q15 format.]

The process of recovering n-bit data from the $2n$-bit accumulator involves an unavoidable loss of information—the low n bits in the case just illustrated. In fixed-point processing, only full amplitude samples get the benefit of the full precision used to specify the coefficient. A better approach, implementable on all DSP devices but with dramatic variations between devices in the overheads involved, is to use a **floating-point** representation of data (cf. Chapter 3). The mantissa is represented in sub-unitary format and the exponent in integer (Q0) format. To perform a multiplication of a pair of signed sub-unitary floating-point numbers the two m-bit mantissas are multiplied, giving a result of length $2m$ bits in Q2 $(m-1)$ format. The lower $m-1$ bits must be truncated before storage, but because both multiplicands have significant bits in the MSB maximum accuracy is retained. The multiplication is completed by adding the two exponents which must be in 2's complement format

Returning to fixed-point processing we have seen that multiplication of a waveform sample with a coefficient gives a double-width result of which only the bits in the high half of the accumulator are retained after storage. This truncation represents a loss of information which can have a dramatic effect in recursive algorithms, causing, for example, instability in certain types of digital filters. If rounding rather than truncation is desired (there is still loss of information), it is necessary to add a 1 to the MSB in the low half of the accumulator before storing the result. Some processors make provision for *automatic rounding*.

To summarise, when multiplying a sub-unitary signed operand by an integer signed operand, the result must be left-shifted one position before the high half of the double-width register is interpreted as the integer result and stored. When multiplying a pair of signed integers, the low half of the result register gives the integer result, *no left shift being needed*. When multiplying a pair of sub-unitary operands, the result must be left-shifted one bit to give the sub-unitary result in the high half. All DSP devices make provision for an optional automatic 1-bit left shift when data is moved out of the high half of the product register for this very purpose.

To find the $Q(n - 1)$ format binary representation for a sub-unitary coefficient, multiply the coefficient by 2^{n-1}, round to get an integer, then convert this integer to binary. If the coefficient is negative, the final step is to take the 2's complement. For example, 0.126 would be represented as a 16-bit 2's complement number in Q15 by the binary equivalent of 4128.768 (0.126×2^{15}) rounded up to 4129 which is

0001000000100001

1.6 DSP device internal sub-systems: hardware description

In this section we examine some of the individual components in the architecture of a DSP microprocessor and make some comparisons between the devices available.

1.6.1 Data memory

Most of the currently available DSP devices have a single data bus which is 16 or 22 bits wide. In the case of the NEC MPD7720 this bus and the associated RAM are internal; in all other cases the data bus and its associated address bus are brought outside to allow memory expansion off-chip. The 32-bit data bus on the MPD77230 and the WE DSP32 allows floating-point storage of data with a 24-bit mantissa and 8-bit exponent. The Motorola DSP56000 and the NEC MPD77220 use a 24-bit-wide data bus, the long word length allowing filter coefficients to be very accurately implemented, thus greatly reducing the chance of internal overflow, despite being fixed point.

Because signal processing operations often involve two real operands, a speed increase can be obtained if an instruction can fetch two data elements simultaneously. Examples are complex numbers (real and imaginary parts), waveform samples and filter coefficients, FFT butterfly inputs, and pairs of samples from signals being correlated. The DSP56000, the 68931 and the MB8764 all have this facility which is provided by having two completely independent data buses.

The DSP56000 can also handle 'long' data words (32 bits) by storing the upper and lower halves in separate data memories from which the two halves may be simultaneously fetched. Similarly, the Thomson 68931 can treat pairs of 16-bit words stored in the location in each memory as a single complex number.

The penalty of having multiple data buses is a very large pin count. In the Motorola DSP56000 a compromise is made by multiplexing the bus for external data memory access. Extra machine cycles are needed for this purpose but the impact can be minimised by judicious memory mapping. The Texas Instruments TMS32020 accesses both external program and external data memory on a single multiplexed bus, possible in this case because instructions and data words are both 16 bits wide.

Some devices without dual memory have provision for using the program memory for data storage. In the Analog Devices ADSP2100 a cache memory is provided to minimise the effect of missed instruction fetches that are a consequence of using program memory. In the TMS32020 a block of internal memory can be software-configured to be either program or data memory; the use of this memory for single-cycle execution of a dual operand fetch-multiply-accumulate operation is illustrated later in this chapter.

Table 1.7 compares the data memory parameters of the various devices. Note that the Motorola DSP56000 and NEC MPD7720 series both have on-chip data (EP)ROM as well as RAM to facilitate coefficient storage. The Texas Instruments TMS320 series can store coefficients in program memory and transfer them to data memory as part of the initialization sequence. Transfers between program and data memory are also supported on the ADSP2100.

It is usual for some data memory locations to be reserved for internal use for such purposes as interrupt vectors and peripheral status registers. Fig 1.7 shows a data memory map of the TMS32020 for each of the internal memory configurations, highlighting the reserved memory locations and their register functions.

Processor	Bus width (bits)	Addressable space (words)	On-chip RAM (words)	On-chip ROM (words)
TMS32010	16	4 K	144	–
TMS32020	16	64 K	544	–
TMS320C25	16	64 K	544	–
TMS320C30	32	16M*	2 K	4 K
NEC MPD7720	16	128	128	512
NEC MPD77230	32	8 K	2 K	1 K
MB8764	16	128 K (×2)	1024	
DSP56000	24	64 K (×2)	256 (×2)	256 (×2)
ADSP2100	16	16 K	–	–
68931	16			
WE DSP16	16	64 K	512	2 K
WE DSP32	32	56 K(bytes)*	4 K(bytes)	2 K(bytes)

*Program and data memory in a single logical unit

Table 1.7 Data memory parameters for common DSP devices

1.6.2   Program memory

The key parameters of program memory are the *instruction width* and the *total addressable space*. The more independent sub-systems there are on-chip, the wider the instruction must be so that bit fields are available for simultaneous directives to each unit. Wide instructions are also needed if the directly addressable range is to be maximised. An alternative approach is to use multiple word instructions. The NEC MPD77230, TMS320C30, AT&T WE DSP32 and Thomson 68931 use a 32-bit width instruction and have the capability for specifying up to eight tasks simultaneously. At the other extreme the TMS32010 uses 16-bit instructions. Table 1.8 summarises the parameters for some common processors.

1.6.3   The ALU, the multiplier and the shifter

With the exception of the Fujitsu MB8764 all the devices incorporate a **multiplier** which, in a single machine cycle, can accept two binary numbers of data bus width, multiply them, and load the product into a double-width result register. The MB8764 has 16-bit input registers but only 26 bits in the result register. Either two input registers can be used or a single register with the second operand being fetched directly from memory during the multiplication instruction cycle. There is always provision for accumulation (addition of multiplicand to accumulator) following multiplication, again within the same cycle. Most devices, including the Motorola DSP56000 and the Analog Devices ADSP2100, use an accumulator register more than

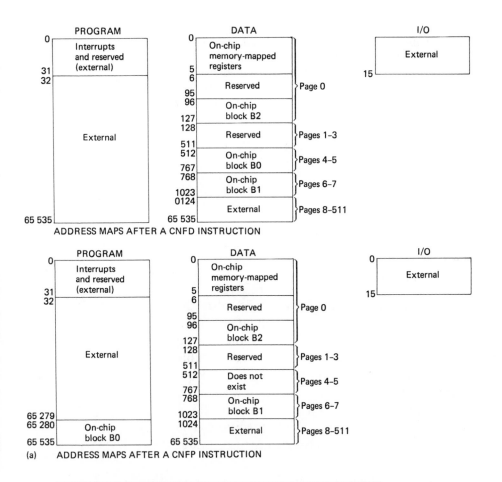

ADDRESS MAPS AFTER A CNFD INSTRUCTION

(a) ADDRESS MAPS AFTER A CNFP INSTRUCTION

Register name	Address location	Definition
DRR(15–0)	0	Serial port data receive register
DXR(15–0)	1	Serial port data transmit register
TIM(15–0)	2	Timer register
PRD(15–0)	3	Period register
IMR(5–0)	4	Interrupt mask register
GREG(7–0)	5	Global memory allocation register

(b)

Fig 1.7 *a*) Texas Instruments TMS32020 memory maps; *b*) Memory-mapped registers

twice the width of the data bus to allow for intermediate overflow during accumulation.

The 32-bit devices have a full floating-point multiplier accepting two 32-bit inputs (24-bit mantissa and 8-bit exponent) to yield a 55-bit result. Numbers can be stored in either floating-point or fixed-point format; in the latter case the 8 exponent bits are zeroed.

In the DSP56000 all arithmetic and logical operations are effected in a single Multiply-Accumulator and Logical Unit. (It is also unique in having a hardware block for bit manipulation of stored data.) On the other devices there is a data ALU separate from the multiplier which performs addition, subtraction and logical operations. Those

Processor	Instruction width (bits)	Addressable space (words)	On-chip memory	Transfer potential
TMS32010	16	4 K	1536	to data memory
TMS32020	16	64 K	256*	to data memory
TMS320C25	16	64 K	4 K + 256*	to data memory
TMS320C30	32	16 M	4 K	same memory
MPD7720	23	512	512	
MPD77230	32	4 K	1 K	
MB8764	24	2 K	1 K	
DSP56000	24	64 K	2 K	
ADSP2100	24	16 K	−	possible
68931	32	64 K	−	
WE DSP16	16	64 K	−	
WE DSP32	32	56 K(bytes)	−	same memory

*On-chip RAM which is software configurable to program memory.

Table 1.8 Program memory parameters for common DSP devices

devices which have two independent data buses (and also the ADSP2100) have two ALU input registers which may be simultaneously loaded. Single data memory devices derive the second ALU input from the accumulator. The accumulator is usually double width, with each half addressable and loadable with sign extension where appropriate (signed numbers loaded into the lower half). Where a single-width accumulator is used there is provision for a carry and a carry input, thus allowing for multiple precision. The ADSP2100 uses this carry facility to support a single instruction (but 16 cycle) divide.

Shift instructions are either implemented in dedicated blocks ahead of and following particular registers, or as a general-purpose **barrel shifter**. The ADSP2100 has independent shifters which can place a 16-bit data word anywhere in a 32-bit-wide result register using logical or arithmetic (with sign extension) shifting as appropriate. Many recent devices support single instruction normalization, that is the scaling of a 2's complement number so that there is a single sign bit, followed by the implied binary point (i.e. the number is always greater than the 0.5). This process enables a fixed-point number to be converted to floating-point—the scaled number is the mantissa, and the number of left shifts required is the exponent. The ADSP2100 can also perform block exponentiation, that is the conversion of a block of input data to a common exponent to make optimum use of available word length.

1.6.4 Address generation

Instructions in DSP microprocessors typically involve the accumulator and data elements in one or two data memory locations. The instruction can incorporate one or both of the data addresses provided they are small enough to fit into the number of bits so allocated in the instruction. This is called *direct addressing*. To reach addresses beyond this range, special registers are provided into which the address can be loaded prior to the instruction execution. This *indirect addressing* technique is particularly useful when blocks of data are being processed since provision is made for automatically incrementing or decrementing the address stored in the register following each reference. Some devices make provision for general arithmetic operations on the addresses stored in these registers during the execution of an instruction. Optimum exploitation of the address generation function for array processing requires careful attention to how the array is mapped onto memory.

The Motorola DSP56000 has extensive address generation capabilities. For each of the two memories, there are three banks of four 16-bit registers, the first holding a current address, the next holding the offset by which the current address should change following each call, and the third a modifier register which determines whether a linear, or modulo *n*, or bit-reversed update should be made. Using these capabilities it is easy to implement *cyclic buffers* (cf. Chapter 3, section 3.6) for input and output as well as more involved address generation for specific algorithms.

An alternative to a cyclic buffer for input signal storage is to dedicate a block of memory for holding the current and previous *n* input samples and to 'shuffle' the data through—each element moving one place forward in each sampling interval. The TMS32010 and 20 series have a Move Data instruction which looks after this task during a multiply-accumulate cycle.

Immediate addressing, that is the generation of constants by incorporating their value into that part of the instruction normally used for an address, is allowed for in certain instructions. In the TMS32020 there is a Multiply Immediate instruction which permits multiplication by a constant coefficient, up to 13 bits resolution, rather than using a value stored in data memory.

1.6.5   Program control logic

Program control performs *instruction pre-fetch*, *instruction decoding* and *exception processing*. The default condition is for instructions to be processed sequentially, accessed by the *program counter* (PC), which is a dedicated register for holding the address of the current instruction. This sequential pattern is suspended when the program branches to execute a conditional statement or loop, or when it is necessary to respond to a peripheral device requiring servicing.

The test for **branching** involves inspection of a special register,

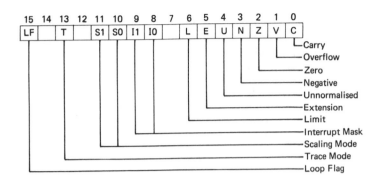

Fig 1.8 Status register in the DSP56000

automatically updated, following arithmetic or logical operations, to flag such conditions as zero, negative result or overflow. Alternatively, branching can be based on the testing of a particular working register. Fig 1.8 gives the significance of the bits in the status register of the Motorola DSP56000. Table 1.9 gives an example of the alternative approach in the form of the set of TMS32020 branch instructions where tests may be applied to the accumulator, the auxiliary register used for indexed addressing, the overflow flag, or the state of an external pin (BIO). Unconditional branches, branches to addresses held in particular registers and subroutine calls are supported on all DSP devices.

In conventional microprocessor architectures the program sequencer responds to **interrupts** or **subroutine calls** by first pushing the current program count value and status register contents onto the stack. The *stack* is a dedicated block of on-chip memory operated on a first-in first-out basis. After saving the context in this way, the program counter is loaded with the address of an interrupt service routine. This routine must itself save and later restore the contents of any of the registers it will rewrite. Once this routine has been completed the previous program count and status register are pulled from the stack and the interrupted process is resumed. High-performance DSP devices have additional hardware to speed up interrupt handling. The DSP56000 can execute single-instruction interrupt service routines without using the program counter. The ADSP2100 has shadow registers backing up the ALU, MAC and shifter input and output registers. The shadow set can be switched in from a subroutine, thus avoiding a context save.

Some DSP devices make provision for the repetition of instructions without the need for software loops and consequent instruction fetch overheads. The more advanced devices make specific provision for loops by providing dedicated loop counter (LC) and loop end address (LA) registers. When a loop is commenced in the DSP56000 processor, the current LC and LA are pushed onto stack followed by

Mnemonic	description	No. cycles	No. words	Opcode instruction register
				15 14 13 12 11 10 9 8 7 6 5 4 3 2 1 0
B	Branch unconditionally	2	2	1 1 1 1 1 0 0 1 0 0 0 0 0 0 0 0
				0 0 0 0 ← branch address →
BANZ	Branch on auxiliary register not zero	2	2	1 1 1 1 0 1 0 0 0 0 0 0 0 0 0 0
				0 0 0 0 ← branch address →
BGEZ	Branch if accumulator ⩾0	2	2	1 1 1 1 1 1 0 1 0 0 0 0 0 0 0 0
				0 0 0 0 ← branch address →
BGZ	Branch if accumulator >0	2	2	1 1 1 1 1 1 0 0 0 0 0 0 0 0 0 0
				0 0 0 0 ← branch address →
BIOZ	Branch on $\overline{BIO}$ = 0	2	2	1 1 1 1 0 1 1 0 0 0 0 0 0 0 0 0
				0 0 0 0 ← branch address →
BLEZ	Branch if accumulator ⩽0	2	2	1 1 1 1 1 0 1 1 0 0 0 0 0 0 0 0
				0 0 0 0 ← branch address →
BLZ	Branch if accumulator <0	2	2	1 1 1 1 1 0 1 0 0 0 0 0 0 0 0 0
				0 0 0 0 ← branch address →
BNZ	Branch if accumulator ≠0	2	2	1 1 1 1 1 1 1 0 0 0 0 0 0 0 0 0
				0 0 0 0 ← branch address →
BV	Branch on overflow	2	2	1 1 1 1 0 1 0 1 0 0 0 0 0 0 0 0
				0 0 0 0 ← branch address →
BZ	Branch if accumulator =0	2	2	1 1 1 1 1 1 1 1 0 0 0 0 0 0 0 0
				0 0 0 0 ← branch address →
CALA	Call subroutine from accumulator	2	1	0 1 1 1 1 1 1 1 1 0 0 0 1 1 0 0
CALL	Call subroutine immediately	2	2	1 1 1 1 1 0 0 0 0 0 0 0 0 0 0 0
				0 0 0 0 ← branch address →
RET	Return from subroutine	2	1	0 1 1 1 1 1 1 1 1 0 0 0 1 1 0 1

Table 1.9 Branch instructions in the TMS32020

the current (loop start) address and the LC loaded with the number of iterations. Each time the end of the loop is detected (PC = LA), the LC is tested, and if not 1, the PC is loaded with the address read (not pulled) from stack and LC is decremented. On exit the stack is *purged* (pulling the top location and discarding its contents) and the previous LC and LA values are restored.

The degree of nesting of loops, subroutines and interrupts depends on the depth of the stack. The TMS32020 has a special instruction for transferring the stack to the top of data memory to provide additional capacity. The ADSP2100 has a dedicated stack for loops, thus making loop and subroutine nesting independent.

1.6.6 Interprocessor communication

DSP devices are often used to perform the signal processing tasks in a system under the overall control of a host processor. Most devices (although not the TMS32010) make provision by means of wait-states, mailboxes, etc., for the sharing of memory between processors and for interprocessor communication. A growing number of processors, including the Motorola DSP56000 and AT&T WE DSP32, are designed with multiprocessor operation in mind, requiring the minimum of external hardware.

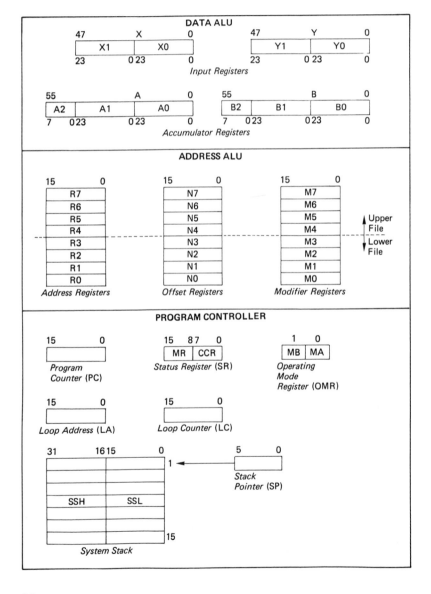

Fig 1.9 DSP56000 programming model

1.7 DSP device software

Figs 1.9 and 1.10 give the internal register set and program and data memory maps for the Motorola DSP56000. The programmer's task is to orchestrate the movement of data between memory and the internal registers in such a way that the ALU operates on the right data samples in the right sequence to perform the desired processing.

Because DSP device architectures allow many processing systems to operate in parallel, conventional sequential-flow high-level languages such as Fortran, Pascal and C are inefficient, certainly when processing speed is critical. The execution time of a particular algorithm will depend as much on the efficiency with which address generation (cf. section 1.6.4) and data movements can be managed as it does on the machine cycle time. DSP software development is very much a matter of fully exploiting the parallel processing capabilities inherent in the architecture and instruction set of the particular device being used.

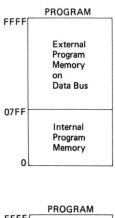
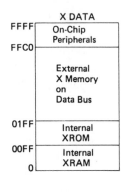
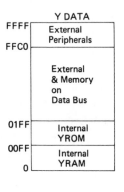

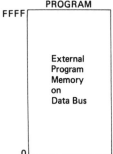
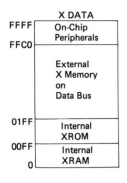

Fig 1.10 DSP56000 memory maps

1.7.1 Instruction building blocks

DSP device instructions can be decomposed into single-action directives belonging to one of the following categories:

Arithmetic operations
Logical operations
Data movement
Program flow control
Condition testing
System control

Arithmetic micro-instructions initiate action by the ALU, multiplier or shifter. Associated with the ALU are Add, Subtract, Absolute Value and Negate. Depending on the architecture there may be a number of variants of addition and subtraction to specify the particular registers involved, the role of the carry bit, how the operand should be scaled, and whether automatic extension of the sign bit is to be invoked. There may be separate instructions for performing arithmetic operations on the auxiliary or address generation registers. Increment, Decrement and Clear are often also provided to facilitate pointer positioning. Associated with the multiplier are the micro-instructions Multiply and Accumulate, again with variants to specify how intermediate overflows are to be handled. Although no device has a single-cycle Divide the more advanced types have an instruction for finding a single bit of the quotient which can be used iteratively for this purpose. Arithmetic operations that can be executed by the shifter in addition to arithmetic (sign extended) Left and Right Shifts for scaling, often include Find Exponent, Normalise and Denormalise operations to facilitate conversion between fixed-point and floating-point formats, as well as an Exponent Adjust instruction.

Logical operations such as And, Or and Complement are supported by the ALU and are used for masking out particular bits of operands, for example the LSBs of signal samples when quantization is to be effected. These instructions can also be used for bit manipulation, although in the more advanced processors specific Bit Manipulation instructions are available which can Test, Set or Clear any bit of a word either in memory or stored in one of the internal registers. Various forms of logical (no sign extension) shift (Left, Right, Rotate Left, Rotate Right) are supported by the shifter.

Data movement covers Loading and Storing of data from and to data memory and also the movement of data between internal registers, between data and program memory (in some processors), and from one location in data memory to another, as is required in a software implementation of a shift register. In processors with dedicated interfaces, FIFOs or ports for data input and output, appropriate Input and Output instructions are available. Where the source and destination registers are of different widths it is necessary to specify the mapping.

Program flow control covers Jumps, subroutine Calls, Returns from subroutines and interrupts, Software Interrupts (a way of invoking the interrupt service routine from the program), Trap and Wait instructions for temporarily halting program execution to allow inter-working with slower external devices, and on the more advanced processors Repeat and Loop instructions for minimizing overheads.

All processors also have a No Operation instruction for implementing software timers and wait states.

Condition testing, when coupled with program flow directives allow the implementation of IF... THEN... ELSE... constructs. Conditions that can be tested vary from processor to processor but usually include the size of the operand relative to zero (zero, negative, positive, etc.), the state of various input lines, and the state of specific flags such as the carry, or overflow, bit in the status word.

Finally, under the umbrella term of **system control** comes a group of instructions which set the mode of operation of the processor and provide various housekeeping functions. Instructions are needed for masking and setting interrupts, pushing and pulling the stack, configuring and resetting peripherals, context saving, and restoring and selecting particular overflow and extension modes. In devices with reconfigurable memory an instruction is needed for allocation of blocks of memory to particular functions.

1.7.2 DSP device instruction sets

As already discussed, DSP devices have instruction words of between 16 and 32 bits width. Each instruction has a number of fields, some of which are used to initiate action by particular processor sub-systems with others being used to designate internal source and/or destination registers and to specify addresses or hold constants. The TMS32020 instruction for loading an auxiliary register with a constant is shown in Fig 1.11.

The 01110 is the code which indicates to the instruction decoder what action is required (the opcode), x specifies which auxiliary register, and y the value of the 8-bit constant to be loaded. In different instructions the bits will be allocated with different formats.

bit	15	14	13	12	11	10	9	8	7	6	5	4	3	2	1	0
	0	1	1	1	0	x	x	x	y	y	y	y	y	y	y	y

Fig 1.11 TMS32020 load auxiliary register with a constant (LARK) instruction

When there is insufficient room in a single word, instructions are sometimes continued over two or more words. This technique is used particularly in devices with 16-bit instructions as it is the only way in which a full 16-bit address or constant can be included in the instruction. In the DSP56000, multiple word instructions are used to provide unrestricted data moves in parallel with arithmetic operations in three *clock cycles* (1.5 machine cycles). The timing of instructions varies from device to device. Some devices, such as the NEC MPD77230 and Analog Devices ADSP2100, adhere strictly to single word instructions which execute in a single machine cycle.

If address information is included in an instruction in the form of a pointer instead of directly, the bits so released can be used to increase the number of simultaneous actions that can be initiated. In devices using wide format instructions and making use of address pointers, a relatively small instruction set is able to support a very powerful parallel processing capability. This is because each instruction can be individually synthesized to perform the precise combination of actions required. In Fig 1.12 the ADSP2100 instruction for any ALU/MAC operation combined with a dual data read (one from program and one from data memory) is given.

23	22	21	20	19	18	17	16	15	14	13	12	11	10	9	8	7	6	5	4	3	2	1	0
1	1	PD		DD		AMF					YOP		XOP			PMI		PMM		DMI		DMM	

Fig 1.12 Analog Devices ADSP2100 ALU/MAC with data and program memory move

The instruction has ten fields defined as follows:

Bits	Purpose
23–22	Opcode
21–20	Register destination of program memory fetch
19–18	Register destination of data memory fetch
17–13	Code for specifying ALU/MAC operation
12–11	Input y register for ALU/MAC
10–8	Input x register for ALU/MAC
7–6	Program memory index register
5–4	Program memory modify register
3–2	Data memory index register
1–0	Data memory modify register

A further example of parallel processing is the fact that almost every instruction (although not the one illustrated) can be made conditional on the state of one of the bits in the status register.

Processors using 16-bit instruction words can also support powerful multifunction instructions; however, each combination of operations has to have its own specific instruction, resulting in large instruction sets and less-readable code. For the TMS32020 there are 17 individual condition tests instructions (Table 1.9), yet the only conditional action supported is branching. Many TMS32020 instructions support combinations of shifts and ALU operations and there are a number of multiply/accumulate/data-move combinations for efficient implementation of summations of products, an often-occurring algorithm in DSP.

1.7.3 DSP software characteristics

A DSP program has two distinctive components. The first handles *signal acquisition*. Every sampling period, one or more signals must be sampled, converted into the appropriate digital format, and stored in the space allocated for this purpose in memory. The second component does the actual *processing*, using as inputs the current signal samples and the results of processing in previous sampling intervals, producing data for external or on-chip peripherals as outputs.

Often the most convenient representation of the process is in the form of a diagram incorporating blocks, each of which represents a delay of one sampling period (labelled z^{-1} for reasons which will become apparent in Chapter 2). Fig 1.13 is an example of a recursive digital filter algorithm represented in this way.

Fig 1.13 Recursive filter representation: signal flow graph for *n*th order filter

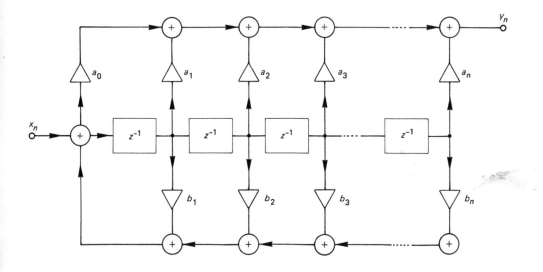

1.7.4 Algorithm description

The algorithm is initially described by means of an equation or flow graph using symbolic names for the inputs and outputs. This form is suitable for documentation and the generation of test data. For our example we will assume that the task of the module is to produce a weighted sum of three stored signal samples with the output set to zero if the result is negative, otherwise passed on without alteration.

Assigning x_0, x_1, x_2 to represent the signal samples, y to represent the output, and c_0, c_1, c_2 to represent the weighting coefficients we have

$$x = x_0 * c_0 + x_1 * c_1 + x_2 * c_2$$
$$y = f(x)$$

where $f(x) = x, \quad x > 0$
$\qquad\quad = 0, \quad x \leqslant 0.$

In documenting the algorithm it is sometimes helpful to further clarify which inputs and outputs are involved by means of a **data flow diagram**. Although trivial for our present example (Fig 1.14) data flow diagrams for processes involving many intermediate stages are much easier to grasp than sets of equations. This will become apparent in Chapter 5 when Fast Fourier Transforms are discussed.

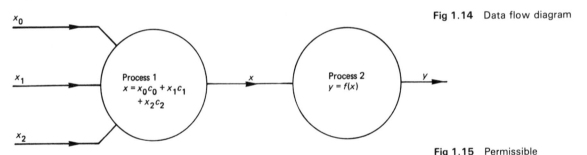

Fig 1.14 Data flow diagram

Fig 1.15 Permissible constructs

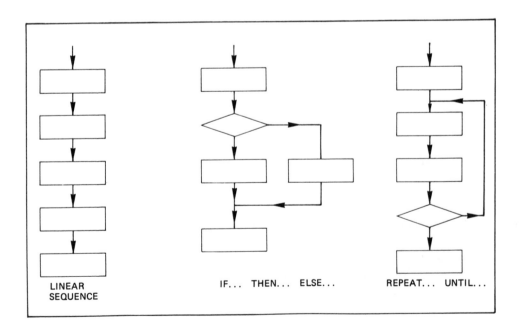

The next stage of the development process is to provide progressively more detail on the sequence of operations that must be performed in order to derive the output from the input. Most processing algorithms proceed through a number of intermediate stages which must be followed in strict order. There are two methods of characterising the sequence of steps in a program, **flowcharts** and **structured English** descriptions. Whatever method is chosen it is good practice to restrict the constructs to those illustrated in Fig 1.15, namely the linear sequence, the conditional branch (if... then... else...), the loop (repeat... until...) and the conditional jump to subroutine (case... of...).

In our three-term weighted sum example the highest-level description of the algorithm in structured English and flowchart format is illustrated in Figs 1.16 and 1.17.

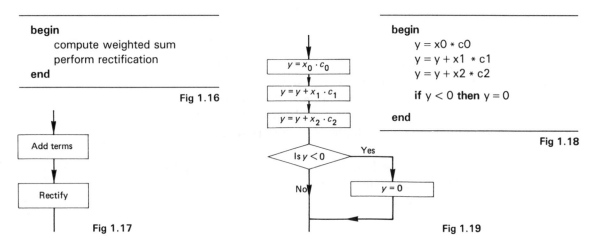

begin
    compute weighted sum
    perform rectification
end

Fig 1.16

Add terms

Rectify

Fig 1.17

$y = x_0 \cdot c_0$

$y = y + x_1 \cdot c_1$

$y = y + x_2 \cdot c_2$

Is $y < 0$ — Yes

No

$y = 0$

Fig 1.19

begin
    y = x0 * c0
    y = y + x1 * c1
    y = y + x2 * c2

    if y < 0 **then** y = 0

end

Fig 1.18

It is usually necessary to expand these high-level descriptions to give more detail, either by implementation of each block as a subroutine with its own internal description or by simply expanding the blocks themselves. While the subroutine approach is normally favoured in DSP programs (programs are then more readable and maintainable), it is a question of balancing the overhead in execution time of subroutine calls against readability and the excess program memory required for repeated sequences when a subroutine approach is not adopted.

In the current example we will simply expand the high-level descriptions as shown in Figs 1.18 and 1.19. Note that at the end of this phase of the development process the algorithm could be directly implemented with a high-level language. It is often profitable to write a program in a high-level language so that the algorithm can be tested off-line using files of simulated input data.

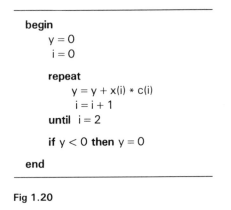

begin
    y = 0
    i = 0

    repeat
        y = y + x(i) * c(i)
        i = i + 1
    until i = 2

    if y < 0 **then** y = 0

end

Fig 1.20

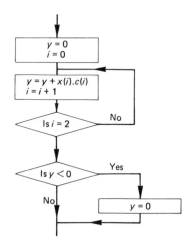

Fig 1.21

There are many possible solutions to the task of flowcharting an algorithm. In our example, an obvious alternative to the linear coding illustrated above would be a loop implementation, with signal samples and coefficients represented as array elements $x(i)$ and $c(i)$, where i runs from 0 to 2. This implementation is illustrated in Figs 1.20 and 1.21.

The decision to use linear or loop coding must be made on its merits considering the overheads in the DSP device to be used and the number of iterations needed.

1.7.5 Writing the assembly language code

Assembly language is one step removed from the binary instruction words actually used by the program decoder. Instructions are specified by mnemonics and internal registers by predefined symbols. Numbers can be represented in hexadecimal, decimal, octal or binary bases or by means of user-defined symbols. The assembler program in the development support computer translates the assembly language code into object code that can be downloaded into the DSP program memory.

Typically each block of the fully expanded flowchart (or each line of the structured English description) results in a line of the assembler source code. Additional lines (instructions) may be needed to provide for tasks such as address generation, mode setting and other housekeeping operations. It is because writing in assembler source code allows the specification of this type of detail (and hence gives the programmer full control of performance) that assembly

32

language programming is required. Otherwise a high-level language compiler could be used to produce object code directly.

Assembly instructions typically have fields separated by spaces for a label, a mnemonic, address information and comments. When parallel operations are supported, additional fields are needed. Fig 1.22 shows the syntax of a Texas Instruments TMS32020 assembler expression.

```
[  <label>  ]    LACK    <constant>        [comment]
```

Fig 1.22 Texas Instruments TMS32020 assembler syntax

```
MR = 0, MX0 = DM(ˆX0), MY0 = PM(ˆC0);
MR = MR + MX0*MY0(SS), MX1 = DM(ˆX1), MY1 = PM(ˆC1);
MR = MR + MX1*MY1(SS), MX2 = DM(ˆX2), MY2 = PM(ˆC2);
MR = MR + MX2*MY2(RND);
AR = MR1;
IF AN  AR = 0;
DM(ˆY) = AR;
```

Fig 1.23 Analog Devices ADSP2100 assembler code for example algorithm (linear)

```
        M0 = 1;
        I0 = ˆX0;
        I1 = ˆC0;
        M1 = 1;

        CNTR = 3;
        MR = 0, MX0 = DM(I0,M0), MY0 = PM(I1,M1);
        DO LOOP UNTIL CE;
        MR = MR + MX0*MY0, MX0 = DM(I0,M0), MY0 = PM(I1,M1);
LOOP    AR = MR1;
```

Fig 1.24 Analog Devices ADSP2100 code for part of example algorithm (looped)

As the actual instruction set and architecture of the DSP device to be used have a major impact on the optimal form of implementation, we illustrate the assembly phase with our example coded for two quite different devices, the 24-bit instruction Analog Devices ADSP2100 (Fig 1.23 and 1.24) and the 16-bit instruction TMS32020 (Figs 1.25 and 1.26). First the complete linear coded version is given, then a section of the code for the loop implementation. In order to facilitate cross-referencing between the flowchart and the assembly listing, the addresses of all variables and constants are given symbolic names that correspond to those used in the flowchart. It is assumed throughout that the signal samples and coefficients have already been loaded into their memory locations.

```
          ZAC                 ZERO ACCUMULATOR
          LT    X2            X2 TO T
          MPY   C2            C2*T TO P
          LTD   X1            X1 TO T, X1 to X2, ACC = P + ACC
          MPY   C1            C1*X1 TO P
          LTD   X0            X0 TO T, X0 TO X1, ACC = P + ACC
          MPY   C0            C0*X0 TO P
          APAC                ACC = P + ACC
          BGEZ  NEXT          BRANCH IF ACC > = 0
          ZAC                 ZERO ACCUMULATOR
   NEXT   SACH  Y,1           LEFT SHIFT AND STORE HIGH ACC IN Y
```

Fig 1.25 Texas Instruments TMS32020 code for example algorithm (linear)

```
   CNFP                CONFIGURE INTERNAL MEMORY
   ZAC                 ZERO ACCUMULATOR
   MPKY   0            ZERO P REGISTER
   LARP   1            POINT TO AUX REG 1
   LARK   AR1,X2       AUX REG 1 HAS X2 ADDRESS
   RPTK   2            LOOP COUNTER = 2
   MACD   C2, * −
```

Fig 1.26 Texas Instruments TMS32020 code for example algorithm (looped)

Explanation of ADSP2100 code The symbols MR (multiply result), MXO, MYO (multiply inputs) and AR (ALU result) represent internal Analog Devices ADSP2100 registers; AN is the negative ALU result bit in the status word, DM and PM stand for data memory and program memory respectively, and $\hat{}$X0, $\hat{}$X1, $\hat{}$X2, $\hat{}$C0, $\hat{}$C1, $\hat{}$C2 and $\hat{}$Y are symbols for the addresses of the input data, output data and coefficients as explained above. The first instruction clears the multiply result register and implements a dual data fetch to load the multiply input registers. Note that the coefficients have been deliberately placed in program memory (PM) whereas the signal samples are in data memory (DM) to make a dual fetch possible since the ADSP2100 has a single data memory bus. The next two instructions are multiply, accumulate and pipelined dual data fetch instructions, the SS signifying that the operands are signed numbers. The next instruction rounds the result after calculating the final product, necessary as the result must be transferred from the 32-bit product accumulator to 16-bit memory. After rounding, the high half is loaded into the ALU result register in order to test for a negative result. The next instruction ensures that if the negative result status flag is set the ALU result register is loaded immediately with zero. The last instruction loads the result into the data memory address corresponding to Y.

When a loop is used (Fig 1.24) the signal samples and coefficients must each be stored in contiguous locations in memory to permit indirect addressing. Before entering the loop, two index registers in the address generator (I0 and I1) are loaded with the start addresses of the samples and coefficients respectively. The modifier registers (M0 and M1) are loaded with +1 so that in each case successive addresses generated are incremented by 1. The register CNTR (the loop counter) is loaded with 3. On entry to each iteration of the loop, the loop counter is automatically decremented by 1; thus after 3 iterations it is zero and the 'counter exhausted' bit in the status word (CE) will be set and the loop exited. Note that indirect addressing, which is obligatory in loops, is desirable in linear code as well since it allows sections of code to be reused.

Explanation of TMS32020 code The TMS32020 cannot normally support dual data fetches. Once the multiplier register (designated the T register) is loaded with the first operand, the second operand can be fetched and the multiplication performed, leaving the result in the 32-bit product [P] register. The code performs the following operations. The first instruction zeroes the accumulator in preparation for the subsequent calculation. The second instruction places the waveform sample X2 into the multiplier T register. The multiply (MPY) command results in the multiplication of the contents of the T register with the stored coefficient C2, and places the result automatically in the P register. The LTD instruction is multifunctional: it causes the T register to be loaded with the second waveform sample, the previous P register result to be added to the accumulator, and the waveform sample to be stored in the next highest data memory address. [Although not needed in the current example this memory data move feature is very useful for implementing a software shift register, as is required in filter implementations (the oldest sample must be stored in the highest address).] The multiplication process is repeated for the next two coefficients. A separate accumulate product (APAC) instruction is used for the final calculation, as no further use is to be made of the T register. A conditional branch statement is used to test if the contents of the accumulator are greater than or equal to zero, saving the accumulator contents (SACH) if this is true, or zeroing the accumulator before saving if not true. The result is stored with a single left shift to eliminate the redundant sign bit. Because the TMS320 series uses 16-bit instructions the address symbols (X0, C0, etc.) are limited to a maximum of 7 bits within the 16-bit word. Any address can be reached however by the appropriate setting of the data page pointer. Note that, unlike the ADSP2100, intermediate overflows cannot be tolerated and there is no automatic rounding.

The looped version of the TMS32020 code makes use of the MACD instruction which supports dual data fetches, one from program memory and one from data memory. In a single instruction, the operands are fetched, the program memory address is auto-matically incremented by 1, the auxiliary register containing the data memory pointer is suitably modified, the multiplication is performed, the product accumulated and data moved to the next highest location in memory. The CNFP instruction configures some of the on-chip RAM as the top of program memory, the bulk of the program memory being external. The coefficients are stored continu-ously in this area with C2 having the lowest address. The auxiliary register (AR1) is loaded with the address of X2 then automatically decremented (∗−) following each call.

When performance is critical the process of implementing an algorithm in assembler source code may proceed through several interactions, with evaluations of the execution time and memory requirements of each implementation in turn.

The above two examples highlight the differences between instruc-tion sets of different processor families. The basic algorithm however is valid for both. Familiarisation with the instruction set of the processor under investigation is thus clearly imperative if optimum performance is to be obtained.

1.7.7 Assembly

Once the algorithm itself has been coded it is necessary to add some directives to the source code to specify the start location of the program in program memory and to define any constants that cannot otherwise be resolved. (Labels for example can be resolved.)

A typical assembler source program for the Texas Instruments TMS32020 is presented in Fig 1.27. Note that on Reset the PC is automatically loaded with zero so this location must specify a jump to the start address (unless zero also). The first few locations in program memory are often associated with interrupts and must have branch instructions to interrupt service routines. A typical program will have an initialization block for setting the mode of operation, initializing the interrupts, loading coefficients into memory, etc., followed by a main program incorporating subroutine calls and finally the subroutines themselves and interrupt handling routines.

The file containing the assembler language source is used as input for the assembler program which typically resides in a host computer (desktop or mainframe). The assembler produces two output files. One is a listing giving the hexadecimal address and instruction codes alongside the source statements. The other is the object file which can be used as input to a simulator program or downloaded directly into program memory [RAM or (EP)ROM].

*Typical assembler source code for TMS 32020. Note that an
*asterisk in the first column denotes a comment line.

Fig 1.27 A complete
assembler source program
for the TMS32020

```
            IDT    'TYP'              :give the program a name

            AORG   0                  :set program code start address

            B      INIT               :send resets to a handler
            B      INTRPT             :send interrupts to a handler
```

*Declare symbols used to reference data memory addresses, and
*constants.

```
ONE         EQU    96                 :to hold the value 1
LIMIT       EQU    97                 :to hold >7FFF

NCON        EQU    2                  :two values to initialise

INPUT       EQU    98                 :to hold input data

ADC         EQU    PA0                :port address of ADC
DAC         EQU    PA0                :port address of DAC

DELAY       EQU    100                :delay count for sampling
```

*Hardware interrupts are redirected to the following interrupt
*handler - this is just a dummy routine in this example.

```
            AORG   >20                :place after vectors

INTRPT      RET                       :return from interrupt
```

*Place a table of data values in program memory to transfer to
*data memory.

```
TABLE       DATA   1,>7FFF            :values for ONE, LIMIT
```

*Resets are redirected to the following initialisation routine.

```
INIT        DINT                      :disable interrupts
            SOVM                      :set overflow mode on
            SSXM                      :set sign extension on
            CNFD                      :block B0 is data memory
            SPM    0                  :no shift on P output
            LDPK   0                  :use data page 0
            LARP   AR1                :use auxiliary register 1

            LARK   AR1,ONE            :set data transfer destination
            RPTK   NCON-1             :set repeat count
            BLKP   TABLE,*+           :transfer from program to data

            CNFP                      :block B0 is program memory
```

Now begin main program.

```
START       IN      INPUT,ADC           :input from ADC

            LAC     INPUT               :load offset binary value
            SUB     ONE,15              :convert to 2's complement
            SACL    INPUT               :save converted value

            LRLK    AR1,DELAY           :load delay counter
PAUSE       NOP                         :padding
            NOP
            BANZ    PAUSE               :delay to give sampling frequency

            LAC     INPUT               :load 2's complement value
            ADD     ONE,15              :convert to offset binary
            SACL    OUTPUT              :save converted value

            OUT     OUTPUT,DAC          :output to DAC

            B       START               :back for next input

            END                         :declare the program end
```

1.7.8   Simulation

The simulator is a program running on the development support computer which accepts an object code input, and under user control simulates, in non-real time, the actions taken by the DSP device as it executes its program. Break points can be inserted or the program single-stepped and the contents of any of the internal registers, program or data memory examined. If trace mode has been set then the history of all internal registers over the previous n machine cycles can also be observed. The user can then optionally change the contents of any register, flag or memory location before resuming the simulated execution.

It is good practice to design a series of test procedures and generate the associated sets of test input data during the program development phase for use on the simulator.

Fig 1.28 gives the screen display of the TMS32020 simulator in its halted state. Some of the menu options are shown in Table 1.10.

```
>>PC =      20        OPCODE = 8210    IN    PREVIOUS PC =      20

              ARP      AR0      AR1      AR2      AR3      AR4    CLOCK
INTEGER        0        0        0        0        0        0       1
HEX            0        0        0        0        0        0       1

                  RPTC             TREG          PREG            ACC
INTEGER            0                0             0               0
HEX                0                0             0               0

>>STK =      10      0      0      0
ST0 : OV = 0   OVM = 0   DP = 4    INTM = 0                    BIO = 1
ST1 : F0 = 0   SXM = 0   XF = 0   CNF = 0   ARB = 0   TC = 0   TXM = 0   PM = 0
INTF0 = 0   INTF1 = 0   INTF2 = 0   TINT = 0   INTM0 = 0   INTM1 = 0   INTM2 = 0   INTM = 0
ENTER INPUT VALUE (IN HEX) OR '' – '' TO RETURN TO MAIN
```

Fig 1.28   Texas Instruments TMS32020 simulator register display

1.7.9   Emulation

The final check on program validity is to load the program into the DSP device program memory and execute on the destination hardware. If an emulator is available (cf. Chapter 7) the user can retain some control over execution by means of break points, etc., and also examine registers as in the simulator. Because the emulation is in real time, using the actual destination system, hardware and timing errors can be identified with a logic analyser.

Command	Function†
BH	Breakpoint Help
C	Continue Simulation
DM or <CR>	Display Main Menu
DT	Display Trace Buffer
EX	Execute Commands from Given File
IOH	Input/Output Help Menu
JF	Select Journal File
L	Load New Object File
MH	Modify/Inspect Memory Help Menu
NB	Set Number of Instructions Until Break (TMS32010)
Q	Quit Simulation
R	Run Simulation
RH	Modify/Inspect Registers/Flags Help Menu
RS	Reset Simulator
SS	Perform Single-Step Execution
ST	Display Register Status
STH	Modify/Inspect Status Register/Pin Help Menu (TMS32020)
STR	Save Trace Buffer
TICH	Interrupt/Timing Help Menu (TMS32020)
TR	Toggle Trace
Z	Zero Clock Counter

† If a command pertains to one processor only, this processor is indicated in parentheses

Table 1.10 Main menu commands for the TMS32020 simulator

1.8 The choice of a DSP device

From the preceding description of current DSP devices, it is evident that the selection of a device suited to a given application is by no means straightforward. Some of the factors which might influence choice are cost, performance (mainly speed), second-sourcing, the likelihood of higher-performance devices with upwards-compatible software becoming available, the upfront investment and development costs, and, most importantly, software and hardware support.

When performance is at a premium the only valid comparison between devices is on an algorithm-implementation basis. Optimal code must be written for both devices and execution times compared. Although devices with wide instruction words, dual memory, instructions with maximum scope for parallel processing, and short

Motorola DSP56000, TMS320C30	Inter-processor communication.
NEC MPD77230, TMS320C30, WE DSP32	Floating-point multiplication.
Texas Instruments TMS32010	Very low cost.
Texas Instruments TMS320E15/E17	On-board EPROM program memory
Texas Instruments TMS32020	Comprehensive software and hardware support, compatible with TMS32010.
Analog Devices ADSP2100	High-level assembler.
Thomson 68931	Complex number multiplication.
AT&T WE DSP16	Low power (0.3 W typical) Fast cycle time (50 ns)
DSP56000, TMS320C30, WE DSP32	On-board peripherals, ease of multiprocessor operation

Table 1.11 DSP device special features

machine cycle times will generally give best performance, it is often the case that the critical factor is the availability of a particular instruction, which if available and fully exploited can make an otherwise unimpressive (and low-cost) device the optimal choice. It is reasonable to say, however, that the more advanced the processor is the less need there will be for exploiting special techniques in order to get the required performance and thus the easier it will be to develop and maintain the associated software.

Apart from speed the other major performance issues are memory size (on-board and externally addressable), ease of handling multiple precision and floating-point arithmetic (in particular, how wide is the data word and what special instructions are provided), and what peripheral devices such as serial and parallel interfaces, timers and multi-processing capability are available on-chip.

Table 1.11 highlights particular areas in which individual DSP devices excel.

2 Theory and application of discrete signal processing

2.1 Introduction

The fact that analog signals can be processed digitally is a consequence of *Nyquist's sampling theorem*. This states that under some circumstances it is possible to completely represent an analog waveform by means of a set of numbers, the numbers corresponding to the level of the signal at regular sampling intervals. This chapter investigates the interface between analog and numerical (discrete) representations of signals with particular emphasis on the theory relevant to DSP system design. Those already familiar with the theory, or just eager to get going, can proceed directly to Chapter 3.

Chapter 2 begins by defining the terms *continuous* and *discrete* signals, and introduces a third signal type, *ideally sampled* signals. Although unrealizable, this latter classification enables the use of continuous system analysis techniques in the prediction of discrete system behaviour.

In the following sections, a brief review of the relationships between time and frequency domain representation of signals is given, followed by the development of a number of simple rules governing the frequency domain effects of typical time domain processes such as sampling, windowing and making a signal periodic. Having established the links between continuous and discrete signals the theory of continuous and discrete linear systems is reviewed, and in particular the characterisation and input/output behaviour of such systems. The appropriate transform techniques for continuous and discrete signal representations are discussed with a view to understanding the extent to which DSP systems can mimic the behaviour of a given analog process. This section is of particular relevance to Chapter 4 which deals with filter design.

The chapter concludes by examining the implications of the foregoing on the feasibility of digital signal processing techniques.

Although mathematics is used to convey a concept clearly and concisely, the aim of the chapter is to give the reader an instinctive 'feel' for the process involved. A good engineer should be capable of understanding the operation and application of a particular system model, without having to grapple with a detailed theoretical analysis. The same philosophy can be applied to digital signal processing.

2.2 Classification of signals

In the following, a number of signal types and associated functions are defined which form an essential part of the theory of digital signal processing. It is well worthwhile taking the trouble to become familiar with these definitions as they will be constantly referred to in subsequent sections and chapters.

2.2.1 Pulse and impulse functions

Three of the most commonly encountered functions are defined in Fig 2.1. The **rectangular pulse** function rect[t] is centred on zero time and has a unit amplitude and width. A rectangular pulse of arbitrary amplitude, width and position can thus be expressed using the rect function as follows:

$$A \cdot \text{rect}[(t - t_0)/\tau]$$

having an amplitude A, a width τ and centred on $t = t_0$.

Similarly the **triangular** function tri[t] is a unit height, unit width triangular pulse. The area under tri[t/τ] is $\tau/2$. If the area is required to be unity, then for a width τ the amplitude must be set to $2/\tau$. If τ is now made to approach zero we get, in the limit, the **unit impulse** function, $\delta(t)$:

$$\delta(t) = \underset{\tau \to 0}{\text{Limit}} \ (2/\tau) \cdot \text{tri}[t/\tau]$$

$$= \underset{\tau \to 0}{\text{Limit}} \ (1/\tau) \cdot \text{rect}[t/\tau]$$

The important properties of the unit impulse (or Dirac) function are that it is zero everywhere except where its argument is zero, at which point it has an infinite amplitude, and that it has unity area, since

$$\int_{-\infty}^{\infty} \delta(t) \cdot dt = 1$$

Note: the signal $A \cdot \delta(t)$ has an area of A, not an amplitude of A.

A is called the *weight* of the impulse. In a graphical representation an impulse should strictly be drawn a standard height, with the weight indicated beside it. However in this book, an impulse will be

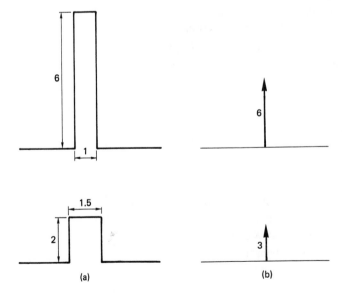

Fig 2.1 Three useful idealised signals

rect(*t*)

tri(*t*)

δ(*t*)

Fig 2.2 *a*) Rectangular pulses with different areas *b*) Impulse representation; the height of the impulse is drawn proportional to its area

6

1

6

1.5

2

3

(a)

(b)

drawn with height proportional to its area since it helps to make diagrammatic representation of sampled signals more explicit (Fig 2.2). (*It is important to recognise that the amplitude of the impulse is actually infinite.*)

45

2.2.2 Continuous signals

Any physical quantity that can be measured, such as temperature, pressure, displacement and current, will exhibit 'smooth' variations with time. It cannot be discontinuous such as the ideal pulse or impulse functions since such changes would imply the absence of any energy storage elements in the associated system.

The term **continuous signal** will be used to refer to any signal that is continuously non-zero over a finite time span. This will encompass all real life signals and also the idealised pulse functions but not the unit impulse function.

2.2.3 Ideally sampled signals

The term **ideally sampled signal** will be used to describe signals that are zero except for discrete (infinitesimally short) moments of time. The impulse function and weighted trains of impulse functions are obvious examples.

2.2.4 Discrete signals

Unlike the above two signal classes which are both functions of time, the independent variable for a **discrete signal** is simply the sequence number. Examples of discrete signals are yearly trade figures, daily rainfall levels or hourly temperature readings. In the first two cases the sequence value represents an aggregate over the sampling interval (a year or a day), whereas in the last, it is a sample of a continuously varying signal, i.e. temperature. The discrete signals encountered in DSP are mostly of this latter type.

In the text, $\{x_n\}$ is used to represent a discrete signal, where the suffix n refers to the nth element in the sequence.

2.2.5 The ideal sampling function

An equally spaced train of unit impulses constitutes an **ideal sampling function** (Fig 2.3). Multiplication of a continuous signal by the sampling function yields an ideally sampled version of the waveform, consisting of a train of impulses whose weights correspond to the instantaneous value of the signal at the sampling moments (Fig 2.4).

2.2.6 Associated signals

Three further operators can be defined which simplify the analysis of DSP systems.

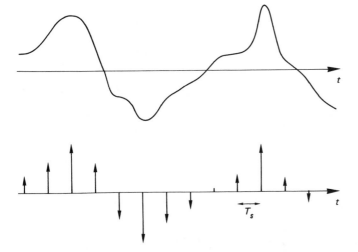

Fig 2.3 The ideal sampling function

1 $\mathrm{per}_{T_p}[x(t)]$

The periodic signal $\mathrm{per}_{T_p}[x(t)]$ is formed by adding together versions of $x(t)$ delayed by nT_p for all integers n, thus

$$\mathrm{per}_{T_p}[x(t)] = \sum_{n=-\infty}^{\infty} x(t - nT_p)$$

This process is illustrated in Fig 2.5*a* and *b*.

2 $\mathrm{sam}_{T_s}[x(t)]$

The ideally sampled signal $\mathrm{sam}_{T_s}[x(t)]$ formed by multiplying $x(t)$ by an ideal sampling function with sampling interval T_s will be represented as

$$\mathrm{sam}_{T_s}[x(t)] = \sum_{n=-\infty}^{\infty} x(t) \cdot \delta(t - nT_s)$$

$$= x(t) \cdot \mathrm{per}_{T_s}[\delta(t)]$$

This operation is illustrated in Fig 2.5*c*.

47

3 $\mathrm{win}_{T_d}[x(t)]$

The operator $\mathrm{win}_{T_d}[x(t)]$ is defined as the multiplication of the input signal $x(t)$ by a rectangular pulse function (window) that has a unit amplitude and width T_d (Fig 2.5d), thus

$$\mathrm{win}_{T_d}[x(t)] = x(t) \cdot \mathrm{rect}[t/T_d]$$

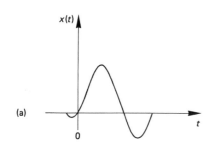

(a)

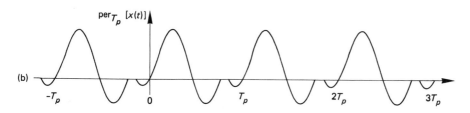

Fig 2.5 *a*) Original continuous signal $x(t)$
  b) Periodic version
  c) Ideally sampled version
  d) Windowed version

(b)

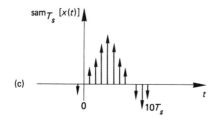

(c)

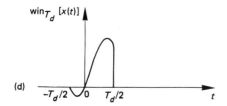

(d)

2.3 The DSP model

In Chapter 1 the basic block diagram of a DSP system was presented. The diagram is reproduced in Fig 2.6*a* with further elaboration in *b* and *c*. Referring to *a*, the antialiasing and reconstruction filters are analog filters whose purpose will be explained later in this chapter. The inputs and outputs of these filters are continuous signals. The A/D and D/A serve to convert the continuous signals into discrete signals and back again, so that at P and Q the signals are discrete.

Fig 2.6 Block diagram of a DSP system

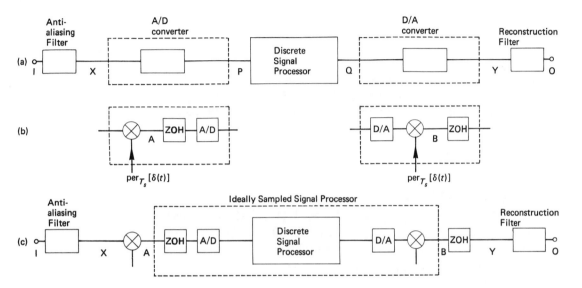

The A/D converter commonly incorporates a *sample-and-hold amplifier*, either internal or external to the actual A/D device, which, when activated, serves to hold the value of the input signal constant, whilst the A/D conversion is performed. The D/A converter similarly holds its output at a constant voltage level corresponding to the digital value in its buffer until the following conversion is effected. Sometimes zeros are output to the converter in between valid words to give output pulses of predetermined width. The 'non-return to zero' and 'return to zero options' are shown in Fig 2.7.

The processor operates on the discrete input signal $\{p_n\}$ and outputs a second discrete signal $\{q_n\}$. Making the assumption that the processor is performing a linear operation on the incoming signal, a convenient means of characterising this operation is in terms of the **unit pulse response**. This is defined as the output sequence $\{h_n\}$ when the input consists of an infinite string of zeros followed by 1 followed by another infinite string of zeros, with $n = 0$ corresponding to the sole non-zero element.

49

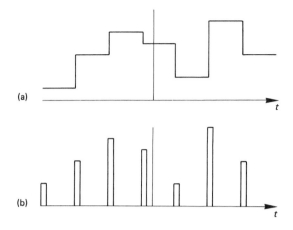

(a)

(b)

Fig 2.7 *a*) Non-return to zero (NRZ) output from D/A converter
b) Return to zero (RZ) output from D/A

Because of the assumed linearity of the processor operation it is valid to deduce the response to an **arbitrary input** by superposition of the scaled and delayed unit pulse responses associated with each term in the actual input sequence, thus

$$q_n = \sum_{i=0}^{\infty} p_{n-i} \cdot h_i$$

An example is given in Fig 2.8. The discrete signal element values are represented by height-scaled vertical lines. Note that the pulse response for a discrete system is the equivalent of the impulse response of a continuous or ideally sampled system. It should not be confused with the continuous pulse function rect[t] that has already been defined.

One of the tasks of the DSP design engineer is to determine the discrete system pulse response that will mimic a specified input/output behaviour (transfer function) between the continuous signals X and Y (Fig 2.6).

A good starting point is to express the input/output relationship for the continuous system in terms of the continuous system impulse response. The relationship is

$$y(t) = h(t) \star x(t)$$

where the $\star$ signifies convolution about which we will have more to say in the next section. This result applies to any time-dependent signals including ideally sampled signals. When the impulse response and the input are both ideally sampled, the convolution reduces to a summation giving

$$b_n = \sum_{i=0}^{\infty} a_{n-i} \cdot h_i$$

Because this is identical to the discrete system input/output relationship above, it suggests that the process of A/D and D/A

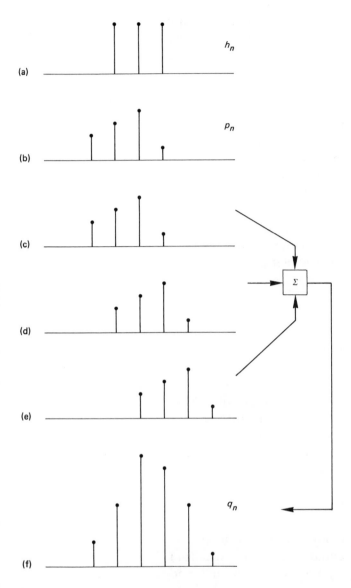

(a)

h_n

(b)

p_n

(c)

(d)

Σ

(e)

(f)

q_n

Fig 2.8 Discrete system input/output derived from pulse response:
  a)   Unit pulse response
  b)   Input samples
  c), *d*), *e*)   Outputs due to individual input samples
  f)   Resultant output samples

conversion can be modelled by the system shown in the dotted blocks (Fig 2.6*b*). Ideally sampled input and output signals are created at A and B by multiplication with ideal sampling functions. The A/D, discrete processor and D/A can all be replaced by an ideally sampled signal processor whose impulse response has impulse weights given by the discrete processor's pulse response.

The overall input/output relationship (between the continuous signals at X and Y) can now be handled by existing continuous system theory in three stages. The first is the effect of multiplying the input by an ideal sampling function. The second is the calculation of the output sampled sequence knowing that the impulse response can be derived directly from the embedded discrete system pulse response. The last step is to determine the effect of the zero order hold process, intrinsic to the D/A converter, on the output sampled signal. The zero order hold has an impulse response of $\text{rect}[(t - \tau/2)/\tau]$ where τ is the width of the actual D/A output pulse. These three stages are addressed in section 2.8.

Summarising, a suitable model for analysis of the discrete part of a DSP system consists of an ideal sampler at the input, an ideally sampled signal processing block whose impulse response consists of a string of impulses with weights equal to the values of the actual discrete system pulse response, and a zero order hold at the output. The analysis model is shown in Fig 2.6c.

At this stage we shall ignore the effects of quantization and rounding error which can be modelled subsequently by adding a suitable noise term to the input signal (cf. Chapter 3).

2.4 Signal analysis tools

2.4.1 Time and frequency domain representations

Most engineers are familiar with signals being represented in the time domain, such as an oscilloscope display. However an alternative method of presenting the same information is the *spectrum domain* or **frequency domain** representation. Provided the signal $x(t)$ exists over a finite period of time, the spectrum can be found by **Fourier transformation**, defined as follows:

$$X(f) = \int_{-\infty}^{\infty} x(t) \cdot \exp(-j2\pi ft) \, dt$$

The function $X(f)$ extends over both positive and negative frequencies and is generally complex. At audio frequencies, spectrum analysers are available which can display both the magnitude and phase of signal spectra in real time—the instruments themselves use DSP techniques which will be discussed in Chapter 5. At RF frequencies, spectrum analysers employ conventional analog processing and it is usually only possible to display magnitude.

The reverse process of deriving a **time signal** from the (complex) spectrum can also be performed using:

$$x(t) = \int_{-\infty}^{\infty} X(f) \cdot \exp(j2\pi ft) \, df$$

The two alternative representations of a signal, one a function of time in seconds and the other of frequency in hertz, are called a *Fourier transform pair*. The following notation is commonly used:

$$x(t) \rightleftharpoons X(f)$$

which implies

$$X(f) = \mathscr{F}[x(t)] \quad \text{and} \quad x(t) = \mathscr{F}^{-1}[X(f)]$$

where $\mathscr{F}[\ ]$ is the *Fourier operator*.

Note the use of upper case for distinguishing the frequency domain $X(f)$ from the time domain $x(t)$.

An apparent limitation of Fourier theory is its inability to handle signals extending over all time. Some signals of this type are **non-stationary**, that is they have time-varying spectra (speech is an example). For this type of signal it makes more sense to divide the signal into blocks of time over which it can be assumed stationary, and find the spectrum of each block. Other infinite-length signals are stationary (such as white noise). For these signals we will show that it is possible to find the spectrum of a *windowed* version of the original time domain waveform which will differ from the true spectrum by an error term which can be controlled by using a large enough window span.

Later we will also give a physical interpretation of the separate magnitude and phase function components of the spectrum and show how certain properties of the Fourier transform can be exploited to avoid having to actually evaluate the integrals in the definitions directly.

The symmetry of the Fourier transform and its inverse are obvious. It would appear that since $X(f)$ is complex, then a complex time domain representation should also be possible and have some physical significance. This is indeed the case. Whereas a *baseband* time domain signal is by definition always real (it is simply the time domain variation of a measured parameter), a **bandpass signal** $g(t)$ of bandwidth $2B$ Hz can be expressed as

$$g(t) = x_i(t) \cdot \cos(2\pi f_c t) - x_q(t) \cdot \sin(2\pi f_c t)$$

where f_c is the carrier frequency and $x_i(t)$ and $x_q(t)$, called the inphase and quadrature components, are both baseband signals of bandwidth B. The spectrum of $g(t)$ can be found by taking the Fourier transform of the **complex baseband** signal:

$$x_i(t) + j x_q(t)$$

(which is called the complex envelope) and translating it from being centred at zero frequency to the frequency f_c and, after frequency reversal and conjugation, to the frequency $-f_c$. The process is illustrated in Fig 2.9. It has great significance in the use of DSP techniques for RF signal processing since it allows an RF signal to be mixed down to baseband (a direct conversion receiver) by a pair of

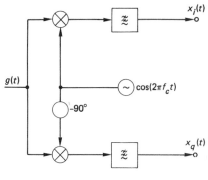

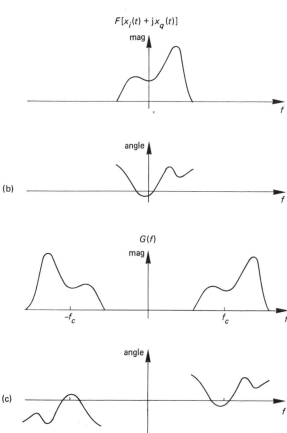

Fig 2.9 Derivation of in-phase and quadrature components of a bandpass signal:
   a) System realization
   b) Spectrum of
$x_i(t) + j x_q(t)$
   c) Spectrum of $g(t)$

quadrature carriers, processed by DSP as a pair of baseband signals, then returned to RF by quadrature amplitude modulation (cf. Chapter 6, section 6.5).

Whether $x(t)$ is complex or purely real is reflected in the symmetry of $X(f)$. For real $x(t)$, the magnitude of $X(f)$ has even symmetry and the phase has odd symmetry about zero frequency. It is because of this symmetry that spectrum analysers need only output the spectrum for positive frequencies.

54

In what follows we will assume that $x(t)$ is a voltage measured across 1 ohm. It is then possible to derive expressions for energy and power without explicitly stating the resistance. $X(f)$ has units of volt-seconds.

2.4.2 Signal energy and the autocorrelation function

Recall that we have restricted $x(t)$ to exist over a finite time interval, and, consequently, provided that the waveform is continuous it has associated with it a finite energy E, which can be derived from either the time or the frequency domain representations. Using $\star$ to represent the complex conjugate of a variable, the relationships are

$$E = \int_{-\infty}^{\infty} x(\tau) \cdot x\star(\tau) \, d\tau \quad \text{volt}^2\text{-sec}$$

which is simply the area under the square of the magnitude of $x(t)$, and

$$E = \int_{-\infty}^{\infty} X(f) \cdot X\star(f) \, df \quad \text{volt}^2\text{-sec}$$

which is the area under the square of the magnitude of $X(f)$.

The function

$$G(f) = X(f) \cdot X\star(f) = |X(f)|^2$$

is called the *energy spectral density function*, and has units of joules/hertz. The spectral density indicates the way in which the energy of the waveform is distributed in the frequency domain.

The energy of a waveform is in fact just a particular value of the more general *autocorrelation function* $\phi_{xx}(t)$. A comprehensive treatment of the autocorrelation function is given in Appendix 5.

2.4.3 Properties of the Fourier transform

As mentioned earlier it is rarely necessary to evaluate the Fourier transform integral directly. Instead we can make use of certain properties which are easily derived from the transform definitions. Before stating some of these properties we will derive one transform using the integral formula itself. It is the transform of a unit impulse function.

For this case,

$$X(f) = \int_{-\infty}^{\infty} \delta(t) \cdot \exp(-j2\pi ft) \, dt$$

Because the impulse is only non-zero at time zero, the integral reduces to finding the area under an impulse, which is just 1. So

$$\mathcal{F}[\delta(t)] = 1$$

Similarly,

$$\mathcal{F}[1] = \delta(f)$$

In other words an ideally sampled signal existing only at zero time has a flat continuous frequency spectrum, and a dc signal has a discrete frequency spectrum which exists only at zero frequency.

Other useful properties of the Fourier transform are as follows. Note $\mathcal{F}[x(t)] = X(f)$ and $\mathcal{F}^{-1}[X(f)] = x(t)$

Linearity

$$\mathcal{F}[x_1(t) + x_2(t)] = X_1(f) + X_2(f)$$

Shift theorems

$$\mathcal{F}[x(t - t_0)] = X(f) \cdot \exp(-j2\pi f t_0) \qquad \textit{time shift}$$
$$\mathcal{F}^{-1}[X(f - f_0)] = x(t) \cdot \exp(j2\pi f_0 t) \qquad \textit{frequency shift}$$

Differentiation

$$\mathcal{F}[x'(t)] = j2\pi f \cdot X(f)$$

Integration

$$\mathcal{F}\left[\int_{-\infty}^{t} x(t) \cdot dt\right] = (1/j2\pi f) \cdot X(f) + (1/2)X(0) \cdot \delta(f)$$

To illustrate the use of these properties consider the task of finding the spectrum of the function $A \cdot \text{rect}[t/\tau]$ which will be needed later in this section.

First differentiate, to convert to impulses whose transform we know:

$$x'(t) = A \cdot \delta(t + \tau/2) - A \cdot \delta(t - \tau/2)$$

Taking the transform using the shift rule, we get

$$\mathcal{F}[x'(t)] = A \cdot \exp(j\pi f\tau) - A \cdot \exp(-j\pi f\tau)$$
$$= 2jA \cdot \sin(\pi f\tau)$$

Finally, using the integration rule we get

$$\mathcal{F}[x(t)] = \mathcal{F}[A \cdot \text{rect}[t/\tau]] = A \cdot \sin(\pi f\tau)/(\pi f)$$

Hence,

$$\mathcal{F}[A \cdot \text{rect}[t/\tau]] = A\tau \cdot \text{sinc}(f\tau)$$

or $\quad A \cdot \text{rect}[t/\tau] \rightleftharpoons A\tau \cdot \text{sinc}(f\tau)$

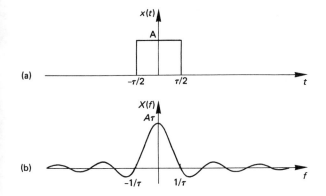

Fig 2.10 Fourier transform
of a rectangular pulse:
a) Time domain
b) Frequency domain

where sinc(x) is defined as $\sin(\pi x)/(\pi x)$. Fig 2.10 gives the time and frequency domain representations of a pulse function.

A useful rule of thumb for estimating the bandwidth of pulse-like signals is that, for a pulse of width τ, 95 percent of the energy is contained in the frequency range from zero to the first zero crossing of the spectral response at the frequency $1/\tau$.

One further transform property we will find useful is that a function evaluated at zero in one domain gives the area under the function in the other domain.

Hence, since $A \cdot \text{rect}[0/\tau] = A$, we can indirectly conclude that

$$\int_{-\infty}^{\infty} A\tau \cdot \text{sinc}(f\tau)\, dt = A$$

i.e. the total area under the curve defined by $A\tau \cdot \text{sinc}(f\tau)$ equals the value of $A \cdot \text{rect}[t/\tau]$ evaluated at $t = 0$.

2.4.4 Some useful transforms

Most of the relationships between associated continuous and discrete time signals can be derived from a knowledge of the transform of the ideal sampling function, i.e. a periodic unit impulse train

$$r(t) = \text{per}_{T_p}[\delta(t)] = \sum_{n=-\infty}^{\infty} \delta(t - nT_p)$$

Repeated application of the time shift theorem gives

$$R(f) = \sum_{n=-\infty}^{\infty} \exp(-j2\pi f n T_p)$$

remembering that $\mathscr{F}[\delta(t)] = 1$. $R(f)$ is obviously periodic along the frequency axis and has an infinite value at $f = \pm k/T_p$ for integer k.

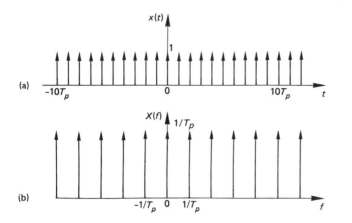

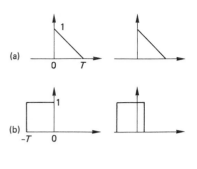

Fig 2.11 Fourier transform
of an impulse train
  a) Time domain
  b) Frequency domain

Fig 2.12 Steps involved in
finding the response of a
linear system with sawtooth
impulse response to a
rectangular pulse input using
convolution.

          a) $h(\tau)$
          b) $x(t-\tau)$
     c) $h(\tau)\cdot x(t-\tau)\,dt$

Using energy arguments we can show that the spectrum is discrete,
consisting of impulses with weight $1/T_p$.

This very important transform pair is illustrated in Fig 2.11. A
train of unit impulses in the time domain (spacing T_p seconds) gives a
train of impulses in the frequency domain (spacing $1/T_p$ hertz), each
with weight $1/T_p$. Equivalently,

$$\sum_{n=-\infty}^{\infty} \delta(t - nT_p) \rightleftharpoons (1/T_p)\sum_{k=-\infty}^{\infty} \delta(f - k/T_p)$$

A time domain ideal sampling function transforms into a frequency
domain ideal sampling function!

2.4.5 Convolution

It is necessary at this point to digress momentarily to consider the
operation of convolution. Suppose $h(t)$ and $x(t)$ are continuous
functions of time. The convolution of $h(t)$ and $x(t)$, represented by

$$y(t) = h(t) \star x(t)$$

is defined as the mathematical operation

$$y(t) = \int_{-\infty}^{\infty} h(\tau) \cdot x(t - \tau)\, d\tau$$

Thus, to evaluate the convolution of two functions at a particular

58

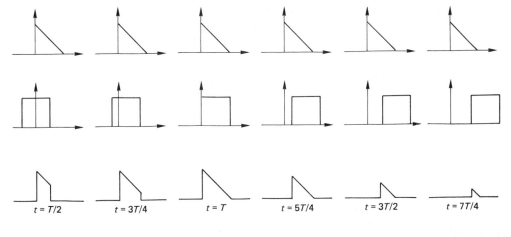

$t = T/2$ $t = 3T/4$ $t = T$ $t = 5T/4$ $t = 3T/2$ $t = 7T/4$

Fig 2.13 Output derived from convolution shown in Fig 2.12

$h(t)$

$T/2$

1/4   1/2   3/4   1   5/4   3/2   7/4   2 t/T

instant in time t, it is necessary to multiply one of the functions with a time-reversed shifted version of the other, and determine the area under the resulting product. This process is illustrated in Fig 2.12 for $x(t)$ being the function $\text{rect}[(t - T/2)/T]$ and $h(t)$ being a sawtooth function of unit height and width T.

Repeating the process for all t and determining the area under the product yields the complete time variation $y(t)$, as shown in Fig 2.13. Note that convolution always involves a smearing out of the waveforms being convolved, the width of the result corresponding to the sum of the function widths. For the case of a unit impulse function, which has zero width, the convolution integral can be simplified to give

$$y(t) = x(t - t_0)$$

assuming $h(t)$ is $\delta(t - t_0)$.

59

In other words, convolution of a function $x(t)$ with an impulse of weight A positioned at $t = t_0$ simply repositions the input function $x(t)$ to the location of the impulse, with scaling A.

Although convolution has been described in the context of the time domain it is of course equally possible to convolve frequency domain functions. For example,

$$A \cdot \text{rect}[f/2B] \star A \cdot \text{rect}[f/2B] = 2B \cdot A^2 \, \text{tri}[f/4B]$$

2.4.6 The convolution multiplication rule

We will make extensive use of the following important Fourier transform property which we state here without proof,

$$\mathcal{F}[x_1(t) \cdot x_2(t)] = X_1(f) \star X_2(f)$$

and

$$\mathcal{F}^{-1}[X_1(f) \cdot X_2(f)] = x_1(t) \star x_2(t)$$

Stated simply, convolution in one domain implies multiplication in the other.

2.5 Windowing and sampling

In this section we use the foregoing analysis to investigate the frequency domain response to three important time domain operations on a waveform, namely making the waveform periodic, sampling the waveform, and windowing the waveform. A set of simple rules is established for the various operations which serve to highlight the main points involved. Because of the symmetry of the forward and inverse Fourier transform definitions, the results can easily be extended to give the time domain consequences of equivalent operations in the frequency domain.

2.5.1 Periodic signals

Given $X(f) \rightleftharpoons x(t)$, what is the Fourier transform of the function $\text{per}_{T_p}[x(t)]$, created by the successive summation of delayed versions of $x(t)$ where the delay is an integer multiple of the period T_p?

The periodic signal is simply $x(t)$ convolved with a train of time domain unit impulses with spacing T_p (cf. section 2.4.5). Using the convolution/multiplication rule we can immediately deduce that the spectrum of the periodic signal is simply $X(f)$ scaled by a factor $1/T_p$ and multiplied by an ideal sampling function in the frequency

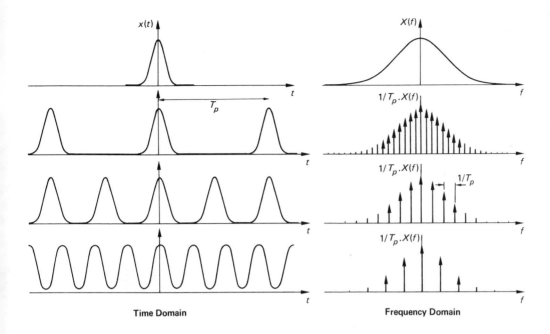

Time Domain   Frequency Domain

domain, the frequency interval being $1/T_p$ hertz. This brings us to Rule 1.

Fig 2.14 The effect on a spectrum of making a one-off signal periodic

---

Rule 1
The periodic version of a reference signal has a spectrum that is an ideally sampled frequency function formed by multiplying the reference spectrum with an ideal sampling function. The scaling factor and sampling interval are both the reciprocal of the repetition period.

The result is illustrated in Fig 2.14 for three cases, two when the original waveform duration is less than the repetition period and one when it is not. We note that a set of numbers—the samples of $X(f)$—are all that is needed to generate the continuous periodic version of $x(t)$. It would appear that there is much redundant information contained in $X(f)$ since we need only retain the value of $X(f)$ at harmonics of $1/T_p$. The original function $x(t)$ is simply a single period of the periodic waveform which results. The only restriction is that the duration of $x(t)$ *must be less* than the repetition period, otherwise the time waveforms overlap and cannot be completely separated. Equally surprising is the fact that many different sets of numbers contain all the information about $x(t)$, i.e. those corresponding to all repetition periods greater than the duration of $x(t)$.

We illustrate this result further by means of an example. Suppose $x(t)$ is one cycle of a cosine wave centred at zero time.

$$x(t) = \text{rect}[t/T_p] \cdot \cos(2\pi t/T_p)$$

or

$$x(t) = \text{rect}[t/T_p] \cdot [\exp(j2\pi t/T_p) + \exp(-j2\pi t/T_p)]/2$$

Using the frequency shift rule and the known transform of $\text{rect}[t/T_p]$, namely the sinc function (section 2.4.3), the Fourier transform of $x(t)$ is given by

$$X(f) = (T_p/2) \cdot \text{sinc}[(f - 1/T_p)T_p] + (T_p/2) \cdot \text{sinc}[(f + 1/T_p)T_p]$$

This function is illustrated in Fig 2.15a. Suppose we now form the periodic waveform

$$\text{per}_{1.5T_p}[x(t)]$$

The spectrum of the periodic waveform is determined by multiplying $X(f)$ by an ideal sampling function with a frequency domain sampling interval of $(2/3T_p)$, and scaling of $(2/3T_p)$. Fig 2.15b illustrates the resulting transform pair.

If we now reduce the repetition period to T_p to coincide exactly with one cycle time of the original sinusoid $x(t)$ then the frequency domain ideal sampling function coincides with zeros in $X(f)$ everywhere except at the frequencies $\pm 1/T_p$. Taking into account the $1/T_p$ scaling, we can deduce that the spectrum of a continuous sinusoid $\cos(2\pi t/T_p)$ is a pair of impulses of weight $1/2$ at $\pm 1/T_p$ (Fig 2.15c).

It is easy to extend this result to show that an arbitrary sinusoid $A \cdot \cos(2\pi f_0 t + \phi)$ has a spectrum with impulses at $+f_0$ and $-f_0$, the *complex* weights being $(A \cos \phi + jA \sin \phi)/2$ and $(A \cos \phi - jA \sin \phi)/2$ respectively. Note that the symmetry in the weights is as we would expect from a real time domain function.

The purpose of the above analysis has been to demonstrate that any periodic function has a spectrum consisting of pairs of impulses at $\pm k/T_p$, where T_p is the repetition period and k is an integer. The pairs of impulses with magnitudes $A/2$ and phases $\pm \phi$ at frequencies $\pm k/T_p$ correspond to sinusoids of the form

$$A \cdot \cos(2\pi kt/T_p + \phi)$$

Hence any periodic signal can be resolved into a summation of sinusoids whose frequencies are harmonics of $1/T_p$, the kth harmonic having an amplitude $(2/T_p)|X(k/T_p)|$ and a phase equal to the angle of $X(k/T_p)$, where $X(f)$ is the Fourier transform of a single period of the waveform centred on zero time. This is in fact the well-known Fourier series representation of a periodic signal, i.e.

$$\mathrm{per}_{T_p}[x(t)] = \sum_{k=-\infty}^{\infty} (2/T_p) \cdot |X(k/T_p)| \cdot \cos(2\pi kt/T_p + \mathrm{ang}[X(k/T_p)])$$

where $X(f) = \mathscr{F}[x(t)]$.

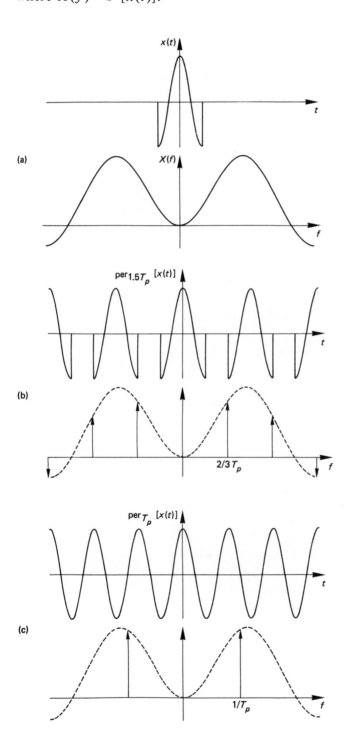

(a)

(b)

(c)

Fig 2.15   *a*) Fourier transform of a single cycle cosinusoid.
   b)   Fourier transform of a periodic version of *a*); period = 1.5 times sinusoid period.
   c)   As for *b*) but with repetition period equal to sinusoid period

2.5.2 Windowed signals

It is often the case in DSP that only part of a waveform can be captured or stored. This process can be modelled as multiplying the true signal with a *rect function* of appropriate duration and is termed **windowing**. Sometimes window functions other than the rect function may be used to achieve a particular effect. This section investigates the relationship between the original and windowed signals in the frequency domain and demonstrates the effect of window shape.

Applying the convolution/multiplication rule, we can show that the Fourier transform of the function

$$\text{win}_{T_d}[x(t)] = \text{rect}[t/T_d] \cdot x(t)$$

is given by

$$\mathscr{F}[\text{win}_{T_d}[x(t)]] = T_d \cdot \text{sinc}[fT_d] \star X(f)$$

which leads us to Rule 2.

Fig 2.16 (*see opposite, p. 65*)
  a) Effect of windowing — wide window:
  (i) and (ii) Original signal in time and frequency domain
  (iii) and (iv) Window in time and frequency domain
  (v) and (vi) Windowed signal in time and frequency domain

  b) Effect of windowing — narrow window;
  (i) and (ii) Original signal in time and frequency domain
  (iii) and (iv) Window in time and frequency domain
  (v) and (vi) Windowed signal in time and frequency domain

Rule 2
The product of a reference signal with a window function has a spectrum that is formed by the convolution of the reference spectrum with the spectrum of the window function.

The process is illustrated in Fig 2.16 where a continuous sinusoid has been windowed firstly with a window duration equal to the sinusoid period and secondly with a much wider window. The effect of windowing is to smear out the spectrum, but as expected, the wider the window the more closely the windowed spectrum approaches that of the original continuous signal. The term **leakage** is often used to refer to the spectral spreading caused by time domain windowing.

The spectral distortion due to windowing can be controlled to some extent by suitable choice of the shape of the window function. For the case of a windowed sinusoid, the extent of the leakage will be less for a triangular-shaped window than for a rectangular window. This is because the spectrum of a triangular pulse, using the fact that

$$(T_d/2) \cdot \text{tri}[t/T_d] = \text{rect}[2t/T_d] \star \text{rect}[2t/T_d]$$

is

$$X_T(f) = (AT_d/2) \cdot \text{sinc}^2[fT_d/2]$$

compared with

$$X_R(f) = AT_d \cdot \text{sinc}[fT_d]$$

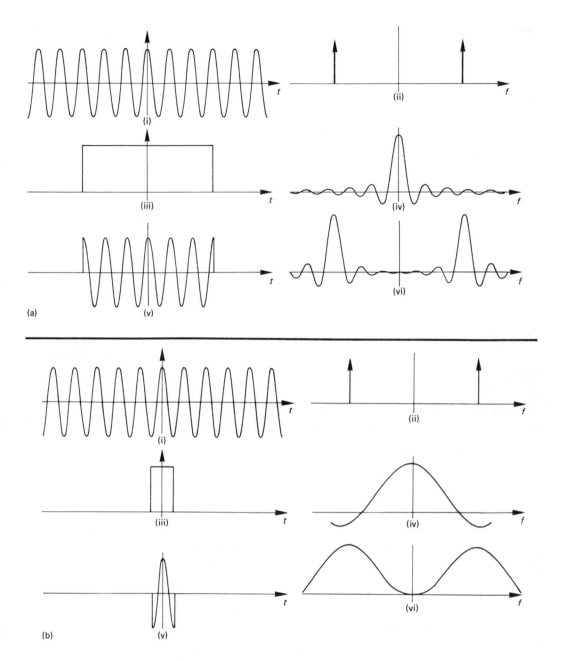

(a)

(b)

for a rectangular pulse, where both have width T_d and height A. The squaring of the sinc function results in smaller spectral components at large frequency offsets, despite the fact that the width of the spectrum at the first zero crossing is twice that of the rectangular pulse. Thus, for a sinusoid with triangular window, the leakage will be worse at frequencies close to the original sinusoid frequency, but better at large frequency offsets, when compared with rectangular windowing (Fig 2.17).

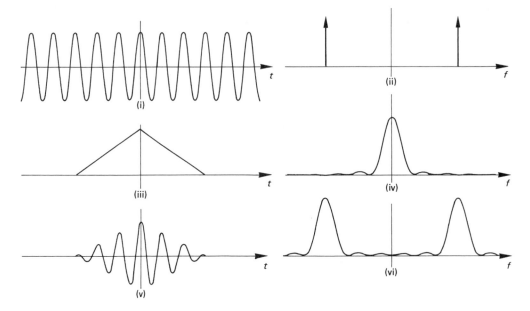

2.5.3 Sampled signals

Fig 2.17 As for Fig 2.16 but with triangular window

If we multiply a continuous signal $x(t)$ by an ideal sampling function we know that the spectrum of the resultant ideally sampled signal will be the convolution of the original spectrum, $X(f)$, with a scaled frequency domain ideal sampling function (Rule 1). Thus

$$\mathscr{F}\,[\text{sam}_{T_s}[x(t)]] = X(f) \star (1/T_s) \cdot \text{per}_{1/T_s}[\delta(f)]$$

where T_s is the time domain sampling interval. The frequency domain ideal sampling function is represented by the periodic train of impulses at multiples of the sampling frequency.

Since convolution of any function with an impulse simply repositions the function at the impulse location (cf. section 2.4.5), we recognise that the spectrum of a sampled signal is just a summation of scaled, frequency shifted versions of the original spectrum, or specifically,

$$\mathscr{F}\,[\text{sam}_{T_s}[x(t)]] = (1/T_s) \cdot \text{per}_{1/T_s}[X(f)]$$

This gives Rule 3.

Rule 3
The spectrum of a signal formed by multiplying a reference continuous signal with an ideal sampling function is a scaled periodic version of the original spectrum. The scaling factor and the frequency domain repetition period are both the reciprocal of the sampling rate.

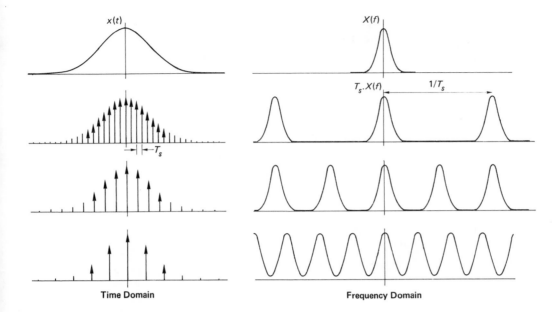

| Time Domain | Frequency Domain |

Fig 2.18 illustrates the above process for two cases. In the first, the repetition period in the frequency domain, $1/T_s$, is greater than the frequency span (twice the bandwidth) occupied by $X(f)$. In the second, the periodicity is less. The results shown are exactly analogous with those obtained in section 2.5.1, but with the domains reversed. This time we find we can throw away information about how the time domain waveform changes between sampling moments, yet still be able to compute the spectrum of the original waveform, provided the sampling rate is high enough.

Fig 2.18 The effect on a spectrum of sampling a one-off time domain signal at various sampling rates.

2.6 Information recovery from sampled signals

In this section techniques are developed for deriving information about a continuous signal from sample values taken at regular intervals in the time domain, i.e. the digital-to-analog conversion process.

2.6.1 Signal recovery: frequency domain

On the basis of Rule 3 the original signal spectrum can be obtained by taking the Fourier transform of the discrete time sampled signal, scaling it by the sampling interval and masking out all repeat images. This leaves the required signal spectrum centred on zero frequency and extending to within half the sampling frequency on either side. It is evident that perfect signal recovery is only possible when the spectrum of the original signal has a total frequency span of less than the sampling rate, or equivalently, that its bandwidth is less than half the sampling rate. This last condition is known as **Nyquist's sampling theorem**. We will call it Rule 4.

> *Rule 4*
> If a continuous signal is multiplied by an ideal sampling function, then the signal can be recovered from the sampled version only if the original signal spectrum is entirely contained within a bandwidth of less than half the sampling rate.

The recovery process is illustrated in Fig 2.19.

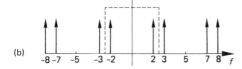

Fig 2.19 *a)* Spectrum of 2 kHz sinusoid
 b) Spectrum of 2 kHz sinusoid sampled at 5 kHz
 c) Spectrum of recovered signal using 'brickwall' low-pass filter at 2.5 kHz

2.6.2 Signal recovery: time domain

To recover the original time domain waveform from its sampled form, the time domain equivalent of the above operation is required. The ideally sampled signal must be convolved with a unit sinc function whose zero crossings occur at the sampling intervals. Convolution with a sinc function is the time domain equivalent of selecting one period in the frequency domain, taking into account the necessary scaling factor. For distortion-free recovery Rule 4 must obviously apply.

We express this result as an equation, and as Rule 5.

$$x(t) = \text{sam}_{T_s}[x(t)] \star \text{sinc}[t/T_s]$$

This rule explains the apparent paradox that the process of sampling appears to discard all the information about the shape of the waveform between sampling moments and yet the sample values still uniquely define the waveform. Note that at the sampling moment itself the only contribution to the recovered waveform is from the sample itself, since those from all other samples correspond to sinc function zero crossings. The process is illustrated in Fig 2.20.

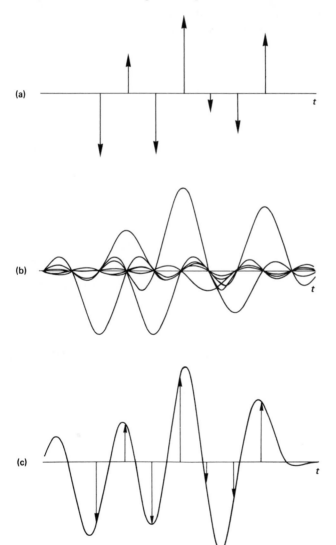

Fig 2.20 *a*) Ideally sampled signal
b) Each impulse in *a*) replaced by sinc (t/T_s)
c) Recovered continuous time waveform

If we attempt to recover a signal which has not been sampled at a fast enough rate, an effect known as **aliasing** occurs. Any arbitrary collection of samples will have associated with it a single continuous waveform whose bandwidth is less than half the sampling rate, and this is the waveform that is recovered by the sinc convolution process. If the samples were generated by sampling a signal with frequency higher than half the sampling rate (i.e. not satisfying the sampling theorem, Rule 4), then the recovered signal will not be the original, but rather the corresponding signal with frequency below half the sampling rate, which would have yielded the same sample values. The in-band component *aliases* a component which actually has a higher frequency. For example, an attempt to recover a 3 kHz sinusoid that had been sampled at 5 kHz would yield a sinusoid of 2 kHz aliasing the true input. A 2 kHz alias would also result if an 8 kHz sinusoid were sampled at 5 kHz.

Figs 2.21 and 2.22 illustrate aliasing in the frequency and time domains. Since alias components represent distortion it is necessary to use an *anti-aliasing* filter prior to sampling.

(a)
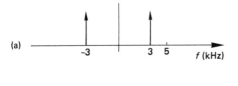

(b)
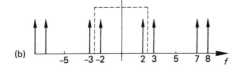

(c)
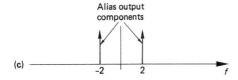

Fig 2.21   Demonstration of aliasing in the frequency domain:
   a)   Original 3 kHz sinusoid
   b)   Spectrum of sampled sinusoid, sampling rate 5 kHz
   c)   Recovered spectrum, with input aliased at 2 kHz

Fig 2.22   Time domain illustration of aliasing. There is only one waveform band-limited to half the sampling rate which fits through the samples. Even though the samples originate from a 3 kHz signal, they are recovered as the 2 kHz signal.

True signal @ 3kHz

Alias signal @ 2kHz

2.6.3 Power and energy

For a continuous waveform of finite duration we have seen that the waveform energy is given by (section 2.4.2)

$$E = \int_{-\infty}^{\infty} |x(t)|^2 \, dt$$

or equivalently,

$$E = \int_{-\infty}^{\infty} |X(f)|^2 \, df$$

If $x(t)$ is made periodic then the total signal energy becomes infinite.

Dividing the energy of a single cycle by the waveform period, the **signal power** is obtained, assuming that the period is greater than the waveform duration. As with energy, the power can also be derived directly from the frequency domain. We know (Rule 1) that

$$\text{per}_{T_p}[x(t)] \rightleftharpoons (1/T_p) \cdot \text{sam}_{1/T_p}[X(f)]$$

$$= \sum_{k=-\infty}^{\infty} X(k/T_p) \cdot \delta(f - k/T_p)$$

As shown above, selecting a single time domain cycle, squaring and integrating will give the waveform energy, which if divided by the period will give the periodic waveform power. The operation of selecting (windowing) one cycle of the time domain is equivalent to convolving the discrete frequency spectrum with a unit sinc function with zero crossings at multiples of the frequency domain repetition period. The component sinc functions in the resulting sum are in fact orthogonal and the energy of the single time domain cycle can be deduced to be

$$E = \sum_{k=-\infty}^{\infty} |X(k/T_p)/T_p|^2 \cdot T_p$$

Hence Rule 6.

Rule 6
The power of a signal, formed by making a reference signal of finite duration periodic, is the sum of the squares of the impulse weights of the periodic signal's discrete frequency spectrum provided the period is greater than the reference signal duration.

The above result could equally well have been derived from the Fourier series representation of the reference signal.

A similar argument can be used to determine the **signal energy** in terms of the time domain samples:

$$E = T_s \sum_{n=-\infty}^{\infty} | x(nT_s) |^2$$

Rule 7
The energy of a reference signal may be deduced from its time domain samples by multiplying the sum of the squares of its time domain samples by the sampling interval provided the sampling rate is high enough.

For a sampled periodic signal, both the time and frequency domain functions are discrete and the power of the original unsampled periodic signal can be deduced from the impulse weights in either domain. We know that the sum of the squares of the time domain samples over one period scaled by the sampling interval gives the energy of one cycle of the unsampled signal (Rule 7).

Similarly from Rule 6, the same result can be found from the sum of the squares of the frequency domain impulse weights over one frequency domain period with suitable scaling by the period (to get energy rather than power) and by the square of the sampling interval (to give the continuous waveform rather than the sampled waveform spectrum amplitude). From this we can state Rule 8.

Rule 8
A sampled periodic signal has a sampled periodic spectrum. The sum, over one period, of the squares of the time domain impulse weights is greater than the sum, over one period, of the squares of the frequency domain impulse weights by a factor equal to the product of the time domain sampling interval and waveform repetition period.

Using a combination of Rules 6, 7 and 8 the energy or power of a reference waveform can be deduced from any of its possible numerical representations.

2.6.4   Correlation and spectral density functions

The autocorrelation function for both energy and periodic power signals was introduced in section 2.4.2. Since $\phi(t)$ and $R(t)$ are just generalisations of energy and power, the results of Rules 7 and 8 can be extended to show how the samples of the original function can be used to derive the samples of the autocorrelation function and spectral density functions.

From Rule 7

$$\phi(nT_s) = T_s \sum_{i=-\infty}^{\infty} x_{i+n} \cdot x_i$$

and hence

$$R(nT_s) = (T_s/T_p) \sum_{\text{period}} x_{i+n} \cdot x_i$$

Taking the Fourier transform of the ideally sampled autocorrelation function

$$\text{sam}_{T_s}[\phi(t)]$$

gives a periodic version of $G(f)$, scaled by $1/T_s$.

Taking the Fourier transform of the ideally sampled power autocorrelation function over one period gives a frequency domain function which can be converted to $S(f)$ by ideally sampling with frequency interval $1/T_p$, scaling by T_s/T_p and windowing out just one period. $S(f)$ is the *power spectral density function*.

A full tabulation of the relationship between continuous signal correlation and spectral density functions and the equivalent discrete operations on the signal samples is given in Appendix 3.

2.7 Linear system behaviour

Although DSP systems can handle linear and non-linear processing with equal ease, linear applications dominate because of the knowledge inherited from the extensive use of such systems in analog signal processing, particularly in the area of filter design.

This section describes how to determine the output waveform when a known input is applied to a given linear system. Both frequency and time domain methods are considered, but with the emphasis on the steady state rather than the transient solution. Two types of system are analysed: the **linear continuous system**, defined as one which can be modelled by the block diagram in Fig 2.23 (the key elements are integrators and summers), and the **linear discrete time system**, modelled by the block diagram in Fig 2.24. Here the key elements are time delay elements and summers. It will be noted that in both cases, the systems are *causal*, that is the output occurs only after the input has been applied.

All physical networks of linear resistors, capacitors, inductors and active controlled sources satisfy the above definition of a linear continuous system.

2.7.1 Continuous systems

The input-output equation for the system in Fig 2.23 can be derived as follows:

$$y(t) = (a_0 + a_1 D^{-1} + \cdots + a_n D^{-n}) \cdot w(t)$$

$$x(t) = (1 - b_1 D^{-1} - \cdots - b_n D^{-n}) \cdot w(t)$$

Hence,

$$(1 - b_1 D^{-1} - \cdots - b_n D^{-n}) \cdot y(t)$$

$$= (a_0 + a_1 D^{-1} + \cdots + a_n D^{-n}) \cdot x(t)$$

where D is the differential operator (D^{-1} is thus the integral operator).

The easiest way to solve this equation is to take the Fourier transform of both sides, making use of the differentiation and integration rules presented earlier and using the substitution $s = j2\pi f$. This gives,

$$Y(f) = H(s) \cdot X(f)$$

where $H(s)$ is called the **transfer function** with

$$H(s) = \frac{a_0 + a_1 s^{-1} + \cdots + a_n s^{-n}}{1 - b_1 s^{-1} - \cdots - b_n s^{-n}}$$

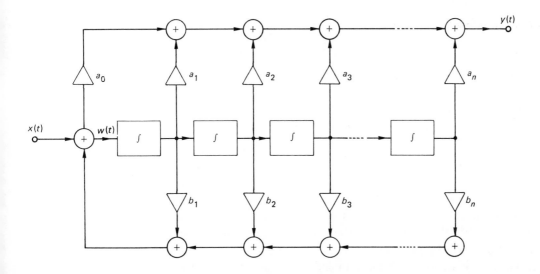

Fig 2.23 Generalised block diagram for any *n*th order linear continuous system.

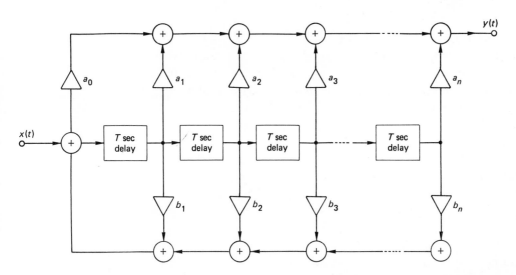

Fig 2.24 Generalised block diagram for any *n*th order linear discrete time system

or equivalently,

$$H(s) = \frac{a_0 s^n + a_1 s^{n-1} + \cdots + a_n}{s^n - b_1 s^{n-1} - \cdots - b_n}$$

This leads to Rule 9 (see p. 76).

A graphical means of representing the transfer function of a network which makes its properties more explicit is to plot the poles and zeros of the function in the s-plane. The **poles** are the roots of the denominator polynomial of $H(s)$, and the **zeros** are the roots of the numerator polynomial. In general the roots will be complex. The s-plane plots the root with the abscissa (x-axis) giving the real part, and the ordinate (y-axis), the imaginary part. Using p and z to represent poles and zeros respectively we have:

$$H(s) = \frac{a_0(s - z_1)(s - z_2)\cdots(s - z_n)}{(s - p_1)(s - p_2)\cdots(s - p_n)}$$

Since the Fourier transform of a sinusoid at frequency f is a pair of impulses at $\pm f$, the response of a continuous system to a sinusoid will itself be a sinusoid, with impulse weights given by the input impulse weights at $\pm f$, multiplied by the transfer function evaluated at f, having complex factors $H(j2\pi f)$ and $H(-j2\pi f)$ respectively.

The gain of the linear system is thus simply

$$\text{Gain} = |H(j2\pi f)|$$

and the phase shift experienced by the sinusoid in passing through the network is

$$\text{Phase shift} = \text{angle } [H(j2\pi f)]$$

Returning to the graphical representation of the continuous transfer function, namely the s-plane pole-zero diagram, it can easily be shown that the complex multiplier $H(j2\pi f)$ can be represented as a product of vectors drawn in the s-plane, one vector being associated with each pole or zero. Each term $(s - x_i)$ in the transfer function expansion corresponds to a vector drawn from the pole or zero location to the point $s = j2\pi f$ on the imaginary axis (Fig 2.25a). The magnitude of the transfer function response $|H(j2\pi f)|$ can be found by taking the product of all such zero vector magnitudes and dividing by the product of the pole vector magnitudes, subject to a constant scaling factor a_0. Similarly the angle of $H(j2\pi f)$ is the sum of the angles of the zero vectors less the sum of the angles of the pole vectors.

If the point on the imaginary axis is swept from zero to infinity (corresponding to the frequency range $0-\infty$ Hz), the *magnitude and phase response plots* associated with the transfer function can be found.

In Fig 2.25a an example is given for the transfer function

$$H(s) = \frac{s}{s^4 - 8s^3 + 94s^2 - 312s + 1655}$$

which has
a zero at $s = 0$
and poles at $s = -1 \pm j6$ and $s = -3 \pm j6$.

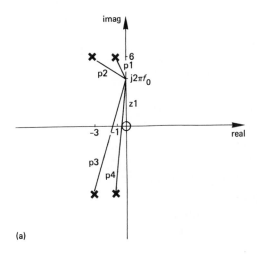

Fig 2.25 s-plane transfer function representation and corresponding magnitude and phase response.

(a)

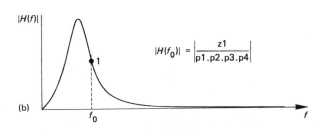

$$|H(f_0)| = \left| \frac{z1}{p1.p2.p3.p4} \right|$$

(b)

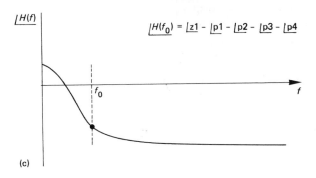

$$\underline{/H(f_0)} = \underline{/z1} - \underline{/p1} - \underline{/p2} - \underline{/p3} - \underline{/p4}$$

(c)

The vectors corresponding to $f = 0.6$ Hz ($j2\pi f = j3.77$) are drawn in. The complete magnitude and phase response are shown in Fig 2.25*b* and *c*.

The response of a linear system can also be expressed in terms of the time domain impulse response. From Rule 9, we know that the Fourier transform of the system output is given by the product of the system transfer function $H(s)$ and the Fourier transform of the input. If the input is a unit impulse, whose transform is unity, then the output response to an impulse is simply the transfer function $H(s)$, and the time domain impulse response of the system, $h(t)$, is the inverse Fourier transform of $H(s)$. The easiest method to derive $h(t)$ is to use a partial fraction expansion of $H(s)$ which, for stable networks, will yield terms which are recognisable as the **Laplace transform** of decaying exponentials. (The Laplace transform is defined as for the Fourier transform but with $j2\pi f$ replaced by s and the lower integration limit changed from $-\infty$ to 0.) For functions such as $h(t)$ which begin at time zero, the Laplace transform is the same as the Fourier transform but with $j2\pi f$ replaced by s. The reason for introducing the Laplace transform at this stage, rather than using the Fourier transform, is because of the less cumbersome expressions that result, and the established body of theory for deducing inverse Laplace transforms without the need to perform an integration. The method will be covered in Chapter 4.

The time domain input-output expression for an input signal other than the impulse is found by taking the inverse Fourier transform of the expression implicit in Rule 9, to yield:

$$\mathscr{F}^{-1}[Y(f)] = \mathscr{F}^{-1}[X(f) \cdot H(f)]$$

or

$$y(t) = x(t) \bigstar h(t)$$

which we encapsulate as Rule 10.

> *Rule 10*
> For a zero state continuous system, the time domain output is the convolution of the system impulse response with the time domain input.

The transfer function $H(s)$ for a continuous linear system can be evaluated in one of two ways: firstly, by direct solution of the network equations in the transform domain (the impedances of resistors, capacitors and inductors being R, $1/Cs$ and sL respectively) and, secondly, by Laplace transformation of the network impulse response if this is known. An example of the first method is given in Chapter 4.

As an example of the second, consider the network whose impulses response is $\text{rect}[(t - T/2)/T]$, i.e. that of a zero order hold system. Using the integration and shift rules, the Laplace transform, corresponding to the transfer function, is

$$[1 - \exp(-sT/2)]/s$$

However derived, the output expression for a continuous linear system will be found to consist of a transient term composed of decaying exponentials or damped oscillations whose time constants are determined by the network poles, and a steady state term which persists beyond the time when the transients have decayed to zero. If the network has stored energy when the input is applied at $t = 0$, the transient term magnitudes and phases will be affected. Techniques based on Laplace transformation of the input/output differential equation are available for finding the complete response, taking initial energy storage into account. A useful rule of thumb is that transients will have decayed to less than 2% of their initial values after 4 time constants, where the time constant is the distance of the pole to the left of the imaginary axis. Networks with positive half plane poles are, of course, unstable.

2.7.2 Discrete time systems

By *discrete time* systems we mean systems whose impulse responses are discrete and take the form of an ideally sampled waveform. The analysis of these systems is exactly analogous to that used for continuous systems in the previous section. This section thus concentrates only on the important results and highlights any significant differences.

The input-output relationship of the discrete time system shown in Fig 2.24 can be found by taking the Fourier transform of the system equation, making use of the shift rules. A less cumbersome notation results if we replace $\exp(j2\pi f T_s)$ by z, giving

$$Y(f) = H(z) \cdot X(f) \qquad (z = \exp(j2\pi f T_s))$$

where

$$H(z) = \frac{a_0 z^n + a_1 z^{n-1} + \cdots + a_n z}{z^n + b_1 z^{n-1} + \cdots + b_n}$$

This leads to Rule 11 (see p. 80).

For an input sinusoid of frequency f, the system output will be a sinusoid of the same frequency with amplitude multiplied by

$$|H(\exp(j2\pi f T_s))|$$

and shifted in phase by $\qquad$ angle $H(\exp(j2\pi f T_s))$

79

> **Rule 11**
>
> For a linear ideally sampled system in the zero state (i.e. no stored energy), the output Fourier transform is the input Fourier transform multiplied by the transfer function $H(z)$ evaluated at $z = \exp(j2\pi fT_s)$, where T_s is the delay.

Graphically the point $z = \exp(j2\pi fT_s)$ is a point on the *unit circle* in the z-plane. As in continuous systems a vector technique can be used for finding the magnitude and phase response. However in this case, increasing frequency is equivalent to the reference point moving around the unit circle. Clearly, the magnitude and phase responses will be repeated with period $1/T_s$ along the frequency axis as this corresponds to one complete revolution in the z-plane. Fig 2.26*a* gives the pole-zero plot in the z-plane for the transfer function

$$H(z) = \frac{z - 1}{z^2 - 1.3856\ z + 0.64}$$

The corresponding magnitude and phase response are given in Fig 2.26*b* and *c* for the range $0 < 2\pi fT_s < 2\pi$.

The impulse response of a discrete time system can be found by dividing the numerator polynomial of $H(z)$ by its denominator polynomial to yield an infinite series of terms of the form:

$$H(z) = h_0 + h_1 z^{-1} + h_2 z^{-2} + \cdots$$

implying

$$H(f) = h_0 + h_1 \exp(-j2\pi fT_s) + h_2 \exp(-j4\pi fT_s) + \cdots$$

and hence

$$h(t) = h_0 + h_1\delta(t - T) + h_2\delta(t - 2T)$$

As for the continuous system, the time domain input-output relationship is

$$y(t) = x(t) \star h(t)$$

However because the system response is an impulse train, the convolution integral reduces to the summation:

$$y(t) = \sum_{i=0}^{\infty} h_i x(t - iT_s)$$

Note that the output $y(t)$ is continuous for a continuous input even though the system impulse response is ideally sampled. The discrete time system can after all be modelled using a tapped delay line.

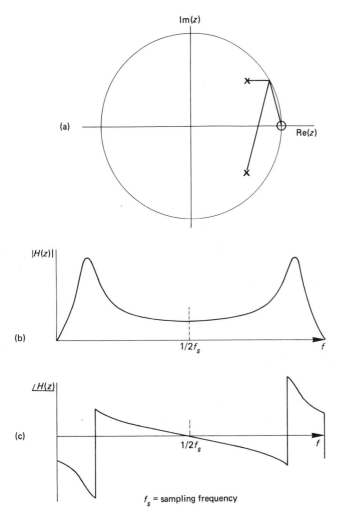

(a)

(b)

$|H(z)|$

$1/2f_s$

f

(c)

$\angle H(z)$

$1/2f_s$

f

f_s = sampling frequency

2.7.3 Discrete systems

Recalling the distinction between ideally sampled and discrete signals, namely that discrete signals correspond to a sequence of numbers which represent the impulse weights of an ideally sampled signal, then the discrete system response $\{y_n\}$ can be expressed as

$$y_n = \sum_{i=0}^{\infty} h_i x_{n-i}$$

where $\{x_n\}$ are the input sample coefficients.

If we consider an ideally sampled input signal,

$$\mathrm{sam}_{T_s}[x(t)] = x_0 + x_1\delta(t - T_s) + x_2\delta(t - 2T_s) + \dots$$

then applying the Fourier transform, we get

81

$$X(z) = x_0 + x_1 z^{-1} + x_2 z^{-2} + \cdots$$

or

$$X(z) = \sum_{n=0}^{\infty} x_n z^{-n}$$

where $z = \exp(j2\pi f T_s)$. This is known as the **z-transform** of $\{x(n)\}$.

The input-output relationship for a discrete system is thus

$$y_n = \sum_{n=0}^{\infty} h_n x_{n-i}$$

or, in the z transform domain,

$$Y(z) = H(z) \cdot X(z)$$

As for discrete time systems, the magnitude and phase response for the sequence can be found graphically from the z-plane pole/zero diagram. Stable discrete systems have their z-plane poles within the unit circle. The transient response decay time is determined by the distance of the poles from the unit circle.

2.8 Application of analysis to the DSP system model

In section 2.3, a model for the complete DSP system, including the analog-to-digital interface was developed which allowed the overall input-output transfer function to be determined. Referring to Fig 2.6c, we know that the relationship between points A and B is given by

$$b_n = \Sigma\, h_n a_{n-i}$$

and the frequency response by

$$B(f) = H(z) \cdot A(f) \qquad (z = \exp(j2\pi f T_s))$$

We also know that since a_n is just an ideal sampled version of the input $x(t)$, and the output $y(t)$ is just b_n after passage through a zero order hold circuit (part of the D/A process), then

$$Y(f) = G(f) \cdot H(z) \cdot (1/T_s)\, \text{per}_{1/T_s} [X(f)]$$

where $G(f)$ is the transfer function of the zero order hold system.

Assuming that no aliasing occurs, i.e. the input frequency does not exceed half the sampling frequency, and that the output reconstruction filter has infinite stop band attenuation for frequencies above half the sampling rate, then the overall transfer function between analog input and analog output (Fig 2.6c) is

$$O(f) = (1/T_s) \cdot I(f) \cdot H(z) \cdot G(f) \qquad (z = \exp(j2\pi f T_s))$$

If steps are taken to compensate for the response $G(f)$ of the zero order hold circuit (cf. Chapter 3, section 3.7.2), then the following system definition can be applied:

> *Rule 12*
> Assuming ideal anti-aliasing and reconstruction filters, a DSP system has an equivalent transfer function determined by evaluating the embedded discrete system transfer function $H(z)$ at $z = \exp(\text{j}2\pi f T_s)$ and scaling the result by $1/T_s$, where T_s is the sampling interval.

To illustrate this result suppose the DSP system has an analog input consisting of a unit amplitude, zero phase sinusoid with frequency 1 Hz. Let the sampling interval be 0.25 second and the impulse response of the discrete process be

$$h_n = 1 \text{ when } n = 0, 1$$
$$= 0 \text{ when } n > 1$$

The system transfer function is thus

$$H(z) = h_0 + h_1 z^{-1} = 1 + z^{-1} = (z + 1)/z$$

which has a zero at -1 and a pole at 0. Evaluating $H(z)$ at the point $z = \exp(\text{j}2\pi \cdot (1) \cdot (0.25))$ gives the amplitude of the output sinusoid as $(1/\sqrt{2})$ and the phase as -45 degrees. Thus, for

$$i(t) = \cos(2\pi \cdot 1 \cdot t)$$

then

$$o(t) = (1/\sqrt{2}) \cdot \cos(2\pi t - \pi/4)$$

2.8.1 The discrete processing algorithm

In designing DSP systems to mimic analog processes we are primarily interested in determining a suitable discrete system transfer function which is equivalent, or nearly so, to the desired analog transfer function $H(s)$. Chapter 4 gives a full coverage of this topic, viewed from the perspective of discrete filter realisations. The outcome of any design process will be a series of feedforward and feedback coefficients (a's and b's) in the generalised discrete system model (Fig 2.24) which yield the correct overall system response $H(z)$. If all the b coefficients are zero in the system model, then the discrete process is termed *non-recursive* or *finite impulse response* (FIR) and has no zeros in the transfer function. In addition, all the product terms generated in the system model are fed forward, hence the term *feedforward coefficients*.

Recursive or *infinite impulse response* (IIR) designs on the other hand may have both feedforward and feedback coefficients, i.e. contain both poles and zeros in the transfer function.

Structures other than that shown in Fig 2.24 can be used to realize a particular system response, examples of which are given in Chapter 4.

2.8.2 Timing jitter

In the theory presented so far it has been assumed that the sampling moments are precisely spaced at a constant interval. In practice phase noise on the sampling clock results in an error τ in the time of occurrence of the sampling moment. It is possible to define the noise due to timing jitter on a particular sample as the difference between the actual sample value and the value that would have pertained had the sampling moment been correct. For a unit amplitude sinusoid it is easy to show that the average **jitter noise** power N_j (assuming that the timing jitter has a flat distribution centred on zero and total width $2J$) is

$$N_j = 2 - \frac{2 \sin (2\pi fJ)}{2\pi fJ}$$

where f is the sinusoid frequency. The corresponding signal-to-noise ratio, using a power series expansion of the sinc function, is

$$\mathrm{SNR}_j = 0.038(T_s/J)^2$$

The signal-to-noise ratio as a function of clock stability is plotted in Fig 2.27.

2.8.3 Quantization noise

So far we have assumed that the signal samples are real numbers. We now examine what effect the quantization introduced by the A/D converter will have on system performance. As is often the case in the analysis of a combined continuous/discrete system, it is convenient to use a model with the same end result but a different internal structure to the actual system in use. In the physical DSP system we have a sample-and-hold amplifier followed by an A/D converter as shown in Fig 2.6b. An equivalent model consists of a discrete time quantization noise source being added to the ideally sampled signal as it emerges from an ideal sampler.

Assuming a uniform amplitude probability distribution to within half the quantization step size Q, the power of the **quantization noise** source is easily shown to be

$$P_{av} = Q^2/12$$

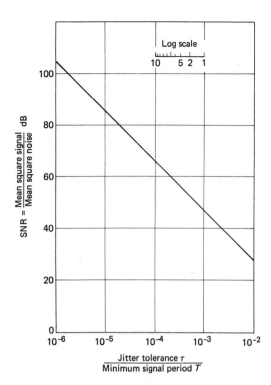

Fig 2.27 Jitter SNR [*Intel 2920 Designer Handbook*]

Log scale

10   5 2   1

SNR = $\dfrac{\text{Mean square signal}}{\text{Mean square noise}}$ dB

$\dfrac{\text{Jitter tolerance } \tau}{\text{Minimum signal period } T}$

If the input signal is noise-like, then the quantization noise power can be assumed to have a uniform spectral distribution over the bandwidth from dc to the half sampling rate. For periodic input signals, particularly where the period is a sub-multiple of the sampling interval, the quantization noise is likely to be concentrated at specific frequencies.

It is possible to reduce quantization noise by sampling the input signal at a very high rate. For example, if the sampling rate is double that dictated by the signal bandwidth, then the quantization noise will be spread over double the signal bandwidth. By digitally filtering the sampled signal to remove components greater than those of the input signal, a 3 dB reduction in quantization noise can be achieved.

A switched variable-gain amplifier ahead of the A/D converter is one method of ensuring that the signal is always as large as possible compared with the quantization noise. A sympathetic switched attenuator is needed following the D/A converter to restore the signal to its former dynamic range. This process is called *companding*. Normally the amplifier gain settings are conveyed by the MSBs of the signal sample digital words.

3 Basic programming considerations

3.1 Introduction

The next four chapters deal with various aspects of algorithm development and application. They cover the basics of data manipulation (using both arithmetic and logical operations), signal processing, waveform generation and filtering/transform techniques.

This chapter is concerned with the very basic application of digital signal processing, providing guidelines for good algorithm and program development technique together with a number of fundamental algorithms which are common to almost all DSP applications. Every effort is made to present the information in a *device-independent* format so that the algorithms should be applicable to any of the DSP integrated circuits currently available on the market. In some cases however, where a number of particular *device-dependent* algorithms exist, the optimal technique for use with the TMS320 series will be given.

3.2 Good design practice

As stressed in Chapter 1, good programming technique plays an essential part in the successful application of DSP. A structured and well-documented approach to programming should be initiated from the outset. It is well worthwhile taking the time to draw up an overall plan of the signal processing task prior to writing any program steps. This plan should take into account the memory requirements, processor-imposed constraints on program length, execution time, etc. The potential use of subroutines for repetitive processes should also be noted, a flow diagram being a very helpful design tool to adopt at this stage (see section 1.7). Where possible, program and data memory blocks should be allocated to specific tasks which optimise data access time and addressing functions (this is particularly important in digital filtering and associated signal processing techniques, Chapter 4). Deciding upon data structures at the outset can save a lot of time and effort later. It is essential to document programs thoroughly with titles and comment statements as this greatly simplifies the inevitable task of troubleshooting and also helps with program familiarisation. *Do not rely on memory*! For ease of understanding, a meaningful choice of mnemonics for variable labels, subroutine names, etc., is to be encouraged. Finally, take full advantage of the hardware and software development support available, e.g. macro libraries, simulation facilities, etc. It is well worthwhile spending some time initially getting to know the development system.

Of particular importance in the successful development of DSP programs is a prior knowledge of suitable algorithms and their requirements in terms of execution time and memory allocation. In many applications, both execution time and memory space are at a premium, the former being dictated by the sampling rate restriction (see section 2.6), and the latter by the particular DSP device. This will often mean a compromise in the type and complexity of algorithm that can be accommodated, and hence the viability of a DSP implementation for a given processing task. It may be, for example, that less accurate but faster executing algorithms must be tolerated in some cases. With experience it is possible to gauge the applicability of DSP for a given application from a block diagram representation of the process in question. To help in this assessment, an indication of processing time, memory requirement, and performance accuracy is given for most of the algorithms presented in this book.

3.3 Initial programming steps

The initial steps of programming involve declaring titles, assigning numeric values to symbols/labels, and declaring data structures. These operations are mainly for the benefit of the assembler itself, and do not result directly in any object code. Following this, any data tables (e.g. look-up tables) to be set up in program memory are declared, and finally the main body of program instructions are entered.

The first operation performed by the processor on power-up or reset should initialize it to the status required by the main program. This involves the setting of interrupt modes, data page pointers, overflow modes and so on. The function of the various modes is described in Chapter 1. Some of the more important are:

Overflow mode How the processor responds to arithmetic overflow in the accumulator.

TMS EXAMPLE No. 3.1

Example of typical initialisation code to set the state of a TMS 32020 after reset, and initialise internal data memory.

```
        IDT     'INIT'              :give the program a name

        AORG    0                   :set program code start address

        B       INIT                :send resets to a handler
        B       INTRPT              :send interrupts to a handler
```

Declare symbols used to reference constants and data memory addresses.

```
PG6     EQU     >300                :start address of data page 6

ONE     EQU     0                   :to hold the value 1
LIMIT   EQU     1                   :to hold the value >7FFF

NCON    EQU     2                   :2 data values to initialise
```

Insert table of data values in program memory.

```
        AORG    >20                 :place after interrupt vectors

TABLE   DATA    1,>7FFF             :values for ONE, LIMIT
```

Hardware interrupts are redirected to the following interrupt handling routine. In this example interrupts are not used, so this is just a dummy routine.

```
INTRPT  RET                         :return from interrupt
```

Resets are redirected to the following initialisation routine.

```
INIT    DINT                        :disable interrupts
        SOVM                        :set overflow mode on
        SSXM                        :set sign extension on
        CNFD                        :block B0 is data memory
        SPM     0                   :no shift on P output
        LDPK    6                   :use data page 6
        LARP    AR1                 :use auxiliary register 1

        LRLK    AR1,ONE+PG6         :set destination for transfer
        RPTK    NCON-1              :number of values to initialise
        BLKP    TABLE,*+            :transfer from program to data

        CNFP                        :block B0 is program memory
```

Now begin main program.

START

Main program code is placed here.

```
        END                         :declare the program end
```

COMMENTS.

The initialisation code shown above is a simple example, for a program which does not utilise interrupts. If interrupts are to be used, then the initialisation code should prepare the interrupt vectors, along with any associated data memory locations.

Data page pointer   Selects the current data page accessed in direct addressing.

Interrupt servicing   Whether the processor responds to a given interrupt vector.

Sign extension   Whether the accumulator treats values as unsigned integers or signed 2's complement numbers.

Auxiliary registers   Define initial working auxiliary register.

A typical block of code which would appear at the beginning of a program for the TMS32020 Processor is shown in Example 3.1.

3.4   Basic processing functions

Having considered the initialization of the DSP device, the next consideration is the mechanisms for data transfer to and from the device, and basic data handling within the device.

3.4.1   Data input/output routines

Interfacing to the outside world for data transfer purposes is one of the most difficult functions to specify. The vast selection of A/D, D/A and logic devices available make it impossible to cater for all application requirements from both the hardware and the software viewpoints. A description of some of the more common hardware peripherals is given in Chapter 7. Whatever the I/O device used, some level of software control will be required for device selection, data buffer enabling/disabling, A/D conversion control, etc.

External input/output device access   The important parameters governing data transfer include the mechanism for selecting and enabling the required input/output device, the initiation of A/D conversion and sample and hold operations for analog input, the means of detecting when valid data is available from input devices and the data input/output transfer itself.

Device selection   In most processors, input and output devices are selected and enabled using a control signal decoded from the processor address bus and a dedicated read/write signal or signals. The task of the software is to set up the correct external device address and activate the input/output control line. This is usually achieved with a single instruction specifying the direction of data transfer, the required address for the external device, and the memory location for storage/retrieval. For example, the TMS32020 assembly language coding for input of parallel data is as follows:

IN STORE,PA6 Input data to memory location STORE
       from I/O port No 6.

This sets the processor's R/W pin high and activates the 2nd and 3rd LSB of the address bus (decoded as port 6).

Analog-to-digital conversion Software control requirements for the A/D conversion process are dictated largely by the external hardware. The important functions are A/D conversion start, and device polling for a valid data or conversion end signal. Both can be accomplished using dedicated software or hardware depending on the application and constraints on program timing, etc. Various techniques are discussed in the chapter on device hardware (Chapter 7).

Resolution/dynamic range Most DSP devices on the market have a 16-bit data bus although some of the more recent devices have 24-bit or 32-bit buses. This limits the data input/output transfer in most cases to a maximum of 16 bits in serial or parallel format, giving a useful signal dynamic range of 96 dB, which is more than adequate for most DSP applications. Where more accuracy is required, additional data bits can be input or output sequentially. In practice, 8 or 12 bit A/D and D/A converters are suitable for the majority of signal processing tasks, with dynamic ranges or 48 dB and 72 dB respectively. Serial input/output devices in the form of codecs, which incorporate analog signal compression/expansion techniques, can be used to obtain increased signal dynamic range. However these generally exhibit slow conversion times and hence require low sampling rates. A more detailed comparison of analog input/output devices can be found in Chapter 7.

3.4.2 Offset binary to two's complement conversion

A common feature of most A/Ds on the market is the use of offset binary for the digital representation of a number, where both positive and negative analog inputs are represented as a binary number ranging from 00000000 for the most negative input to 11111111 for the most positive input, assuming an 8-bit converter. Similarly, the majority of D/A converters accept offset binary input data which may or may not be converted into a bipolar output signal by appropriate hardware. Current DSP devices however use 2's complement representation of numbers in arithmetic operations. Thus, when interfacing to a DSP, software or possibly hardware conversion between offset binary and 2's complement representations must be performed.

If the converted signal is input on the most significant bits of a 16-bit data bus, then the maximum and minimum input values for a 12-bit A/D converter appear in binary form as

$$+ V_{max} = 111111111111\text{XXXX} \qquad (\text{X} = \text{unaffected bits})$$
$$- V_{max} = 000000000000\text{XXXX}$$

Note: LSBs can be either left unaffected, hardwired on a buffered bus, or masked off using a suitable software routine.

To convert from offset binary to 2's complement numerical representation, one of the simplest methods is to subtract a constant value OFFSET, corresponding to half the input voltage swing. With data input to the most significant bits as above, the value OFFSET required is given by

$$\text{OFFSET} = 1000000000000000$$

Binary subtraction of OFFSET from the input data generates the values

$$+ V_{max} = 011111111111\text{XXXX}$$
$$- V_{max} = 100000000000\text{XXXX}$$

which are the desired full-scale 2's complement representation.

Adopting the inverse of the above procedure, i.e. adding the value OFFSET to the 2's complement number, restores the data to an offset binary format for output to a D/A. This process, together with a software masking procedure, is illustrated in Example 3.2.

An alternative procedure, as outlined in section 1.5, is to simply reverse the MSB of the input word (provided it corresponds to the sign bit in the DSP), using an XOR operation with a suitable mask. This method is also illustrated in Example 3.2.

The latter process of bit reversal can equally well be performed in hardware, using a simple logic gate. This has the obvious advantage of reducing software overheads when program memory or execution time is at a premium, but at the expense of increased hardware.

3.4.3   Program timing

The most obvious factor which dictates program execution time is the desired *sampling rate*. This must be at least twice the maximum frequency of the input signal component to avoid aliasing. Other factors such as A/D conversion time or algorithm complexity may also impose restrictions on the maximum sampling rate obtainable, and should be borne in mind when assessing the suitability of digital signal processing for a given application.

TMS EXAMPLE No. 3.2

Illustration of conversion between offset binary and twos complement notation, in association with input from an ADC or output to a DAC.

Initialisation code to set the state of the TMS after reset, and initialise data memory should be inserted here. The data locations HIMASK and OFFSET should be set to >FFF0 and >8000 respectively.

Begin main program.

ADC	EQU	PA0	:use port 0 for adc
DAC	EQU	PA1	:use port 1 for dac

START

Subtract or add offset to convert between offset binary and twos complement.

IN	SIGNAL,ADC	:input from ADC

LAC	SIGNAL	:use accumulator
AND	HIMASK	:discard 4 lsb's
SUB	OFFSET	:subtract offset
SACL	SIGNAL	:save 2's complement result

LAC	SIGNAL	:use accumulator
ADD	OFFSET	:add offset
SACL	SIGNAL	:save offset binary result

OUT	SIGNAL,DAC	:output result

Alternative method for converting between offset binary and twos complement – invert sign bit (msb).

IN	SIGNAL,ADC	:input from ADC

LAC	SIGNAL	:use accumulator
AND	HIMASK	:discard 4 lsb's
XOR	OFFSET	:invert sign bit
SACL	SIGNAL	:save 2's complement result

LAC	SIGNAL	:use accumulator
XOR	OFFSET	:invert sign bit
SACL	SIGNAL	:save offset binary result

```
        OUT    SIGNAL,DAC          :output result to DAC


        B      START               :back around loop

        END                        :declare end of program
```

PROGRAM DETAILS:

Instruction cycles	6 for input and adjust, 5 for readjust and output.
Program memory words	9 for either method, including I/O. (Not including initialisation and loop code.)
Data memory words	3 for either method.

COMMENTS.

Data memory used to store the two masks (HIMASK and OFFSET) could be released by the use of 'immediates' instead. This would impose a time penalty of one extra instruction cycle for each replacement.

If a data memory location holding the value 1 is already available (often the case), then this could be used, with a left shift of 15, in place of the offset mask.

In many devices, a precision timing reference is made available which can be software controlled to cause an interrupt for initialization of a single program execution cycle. The interrupt repetition rate thus determines the program sampling rate. This technique for setting the sampling interval has the attraction of making the overall program timing independent of both processing algorithm and external device timings, provided that input/output conversion routines can all be completed between interrupts. If no suitable 'on-board' clock is available, external clocking and interrupt timing can usually be applied, as outlined in Chapter 7.

In cases where a system clock and interrupt timing mechanism are not available or desirable, the sampling rate can be controlled by the program execution time. At the end of each sampling period, the program loops back to begin processing the next sample, so that the length of the program loop defines the sampling rate. A basic time delay loop can be easily implemented within any program to ensure a given execution rate (see Example 3.3). However this overall approach can make program modification difficult and time consuming, with continual readjustment of the delay required to compensate for any resulting increase or decrease in program execution time.

3.4.4 Memory management

A facility exists on most modern DSP devices for manipulation of data blocks within internal and external memory. This allows large volumes of data or program code to be stored off-chip in slow access memory and transferred to fast-access on-chip memory when required. The mechanisms involved are largely dependent on the chip architecture and reference to the manufacturer's operating manual is essential to determine optimum memory management procedures for a given device.

TMS EXAMPLE No. 3.3

Example of sampling rate control, using a software delay loop

The example implements an input and output sampling rate of 8 kilohertz, assuming an instruction cycle time of 200 nanoseconds for a TMS 32020.

Now begin main program.

ADC	EQU	PA0	:use port 0 for adc input
DAC	EQU	PA1	:use port 1 for dac output
DEL	EQU	153	:for loading delay loop counter
START	IN	SIGNAL,ADC	:input from adc

*2 (count of instruction cycles so far)

	OUT	DAC,SIGNAL	:output to dac

*4

	LRLK	AR1,DEL	:load delay counter
	NOP		:padding

*7

PAUSE	NOP		:pad out delay loop
	NOP		
	BANZ	PAUSE	:decrement count and loop

*623

	B	START	:back for next input

*625 (number of instruction cycles in 8 kHz loop)

	END		:declare the program end

COMMENTS.

For delay loops of 257 cycles or less, it would be more efficient to use repeated NOP's, utilising the repeat counter (RPTK).

3.5 Basic DSP operations

This section covers the fundamental functions which can be performed by modern DSP devices, many of which are common to general-purpose microprocessors, with the exception of the fast multiply facility from which stems a number of unique DSP attributes outlined in this and subsequent sections.

3.5.1 Logical operations

All DSP devices have a command set for performing logical operations, such as AND, OR, XOR, etc. These operate in a manner analogous to discrete logic gates. Their main use is for masking wanted or unwanted data bits, and in bit testing routines, although a number of processors have built-in bit-test commands in addition to the conventional logical functions. This facility is used frequently in floating-point arithmetic, curve fitting, and a number of the more advanced algorithms described in subsequent sections.

3.5.2 Arithmetic operations

The basic arithmetic operations performed by current DSP devices are described in this section. The simple operations of addition, subtraction, multiplication, etc., are provided as implicit single-cycle instructions in all processors and are thus not discussed in detail except where additional specialised commands are utilised in the more involved arithmetic operations described later. Tasks such as binary division may involve a number of instruction cycles, depending on the accuracy of the quotient desired.

Multiplication All DSP devices have a dedicated hardware multiplier which performs the binary multiplication of two data words (typically 16-bits wide) in a single clock cycle. This 'on-board' *high-speed* dedicated hardware multiplier is the one main feature which distinguishes DSP devices from general-purpose microprocessors, and forms the heart of most signal processing algorithms. The speed of 16-bit multiplication along with the associated data manipulation is a primary factor in determining the overall efficiency of the DSP device. A number of dedicated multiply/accumulate/data move commands are built into most processors specifically for filtering and FFT-based algorithms. These take full advantage of the parallel architecture and pipelining of modern processors (see section 1.4) to achieve very rapid program execution. As many commands are application and algorithm dependent, these will be covered in subsequent chapters as they arise. An illustration

96

TMS EXAMPLE No. 3.4

Simple multiplication example.

```
START     LT      VALUE1          :put first operand in T reg.
          MPY     VALUE2          :multiply with second
          PAC                     :load accumulator from P reg.
          SACH    RESULT,1        :lose extra sign bit and save
```

COMMENTS.

        Care must be taken when saving and shifting the result of a multiplication, so that magnitude bits do not replace the final sign bit. (see the section on overflow).

        For multiplication operations, e.g in filtering, the MAC and MACD instructions can be used in repeated mode, and are more efficient for filters longer than about 4 taps.

of the basic multiply routine utilised in the TMS320 processors is given in Example 3.4. Because of the internal architecture of the processor, one of the operands (data words) has to be loaded into the T register of the device which feeds the dedicated multiplier. The product is automatically placed in the P register, the contents of which must then be transferred to the accumulator. A number of options are available such as, APAC (add P register to accumulator), SPAC (subtract P register from accumulator), etc. Some of the advanced DSP devices with enhanced parallel architectures allow a number of the stages outlined above to be executed within one clock cycle.

Division   The division of two 16-bit numbers is considerably more complex than the multiply, add or subtract operations. The algorithm involves a repeated series of shift and conditional subtract operations. The more modern processors have an implicit division command as part of their instruction set which executes this algorithm directly. The TMS320-20 among others, however, requires the algorithm to be implemented 'discretely' and it is therefore described in some detail in the following. In a few specialised applications, the operation of division is better handled by the use of curve fitting routines, and the reader is referred to this section in the chapter for further information.

   The division algorithm for binary arithmetic is implemented in much the same way as for long division using decimal arithmetic. The process is illustrated by way of an example using base 10 arithmetic as follows.

Consider the calculation

$$40/3 = 13.3333$$

Rather than performing this actual calculation we shall compute the result of

$$40/30 = 1.33333$$

bearing in mind that to obtain the correct result we need to multiply the answer by 10. This factor of 10 is introduced to ensure that, at the start of the division, the numerator is no greater than ten times the denominator as the algorithm is unable to cope with a result greater than 9.999... (The scaling of 10 corresponds to the base 10 arithmetic being used.)

The calculation then proceeds as follows:

		Result
30/ 40		
30		1.
100		
90		1.3
100		
90		1.33
100		
90		1.333
10		
	etc.	

The first step is to determine the integer number of times the denominator [30] is subtractable from the numerator [40] giving a positive or zero remainder. In this case it is [1] leaving a remainder of [10]. The value [1] is thus taken as the most significant digit (MSD) of the result. The process is then repeated having mentally multiplied the remainder value by 10, i.e. [100], giving an answer of [3] and a remainder of [10]. This result must then be mentally divided by 10 and represents the first digit after the decimal point, i.e. [0.3]. Again, multiplying the remainder by 10 (100 overall), the result is [3] and the remainder [10]. Repeating the process for the final stage gives an answer of [3] and a remainder of [10]. Thus the final quotient is [1.333]. We must remember to restore the factor of 10 introduced at the outset to give [40/3 = 13.33].

An identical procedure is used in binary arithmetic, the calculation being performed using powers of two. Consider the same example:

$$40/3 = 13.3333$$

As for the decimal arithmetic example, the first step is to introduce a number of left shifts in the denominator to ensure that the numerator is less than twice (since the arithmetic base is now 2) the denominator. For this example, a shift of 3, corresponding to a multiplication factor of 8, is required and the calculation to be performed becomes

$$40/24 = 1.6666$$

Using 8-bit 2's complement binary arithmetic and working with the upper half of a 16-bit accumulator the division proceeds as follows:

(1)   00101000   00000000   ($= 40$)
       $-$ 00011000   ($= 24$)
       ————————————
       00010000   00000000   ($+$ve result)   Answer $= 1$.

(2)   00100000   00000000
       $-$ 00011000
       ————————————
       00001000   00000000   ($+$ve result)   Answer $= 1.1$

(3)   00010000   00000000
       $-$ 00011000
       ————————————
       11111000   00000000   ($-$ve result)   Answer $= 1.10$

(4)   00100000   00000000
       $-$ 00011000
       ————————————
       00001000   00000000   ($+$ve result)   Answer $= 1.101$
                         etc.

The result of step 1 is positive; thus the first bit in the answer register is set to 1, and the remainder from the calculation is 'mentally' multiplied by 2 (base 2 arithmetic) and used as the numerator for the second stage of the calculation. The result of step 2 is also positive, so the second bit in the answer register is set, the remainder multiplied by 2, and the process repeated. The third stage produces a negative result which means that the third bit in the answer register remains unset, the result is discarded, and the current numerator multiplied by 2. The fourth step gives a positive result so the fourth answer bit is set and the remainder from the calculation multiplied by 2 and placed in the accumulator.

     The algorithm can be repeated indefinitely to increase precision. However the word length in the processor limits the length and hence accuracy of answer that can be stored.

The binary result obtained after eight cycles is [1.1010101] and represents the desired quotient 1.664 if the implied decimal point occurs directly after the MSB of the answer as shown. This will in fact always give the correct scaling provided that the numerator is always less than twice the denominator prior to the calculation. With an appropriate right shift of three places, the real quotient is obtained [1101.0101], with a value of 13.31.

The algorithm as shown requires the answer derived at each step to be stored in a separate register. This requirement can be overcome by exploiting a useful property of the long division process, namely that if the answer from each stage, i.e. 0 or 1, is inserted into the LSB of the accumulator (numerator) used in the subsequent stage, it does not affect the result of the calculation. This property is demonstrated in the following identical division example:

(1)   00101000 00000000 (= 40)
        − 00011000   (= 24)
        ─────────────────────
         00010000 00000000   (+ ve, hence shift result and add 1)

(2)   00100000 0000000*1*
        − 00011000
        ─────────────────────
         00001000 00000001   (+ ve, hence shift result and add 1)

(3)   00010000 000000*11*
        − 00011000

        ─────────────────────
         11111000 00000011   (− ve, shift numerator by 1)

(4)   00100000 00000*110*
        − 00011000
        ─────────────────────
         00001000 00000110   (+ ve, hence shift result and add 1)

         00010000 0000*1101* (= ACCUMULATOR)
                      etc.

After a full eight steps, corresponding to the number of bits in the data word, the result held in the accumulator will be

ACCUMULATOR = 00010000   *11010101*

In this case, the ACCUMULATOR register consists of two parts, a QUOTIENT term [11010101] in the *lower* half of the accumulator register and a REMAINDER term [00010000] in the *upper* half of the accumulator. Due to the nature of the algorithm, these values require scaling by a factor of 0.5 (right shift of 1) to give the correct result.

Compensating for the original left shift of 3 at the same time gives the real quotient [00001101 = 13] with a remainder of [00000001 = 1].

Note: The above calculation is only valid if the number of algorithm execution cycles does not exceed the data word length N (i.e. 8 in this example). If greater accuracy is required than can be achieved with N calculations, then the current answer can be stored, and the algorithm repeated to achieve greater accuracy.

The shift, or scaling, imposed on the denominator or numerator prior to the execution of the division algorithm affects the overall accuracy of the result for a given number of calculations. The scaling must be such that the numerator is initially always less than twice the denominator but, for greatest accuracy, preferably still greater than the denominator. If lower values of scaled numerator are used, then the first few bits in the answer register are zero, resulting in reduced accuracy for the final result. *Note: Always make sure to remove the scaling factor by suitable shifting of the final remainder* (ACCUMULATOR).

Implementation The optimal implementation of the division algorithm is device dependent. Most devices have a specialised instruction or instructions which perform a number of the steps in the algorithm in one cycle. For example, the TMS32020 has a conditional subtract command, SUBC, which performs the following functions:

1) Subtract denominator from numerator (held in ACCUMULATOR).
2) If result is negative, multiply numerator by 2 (shift ACCUMULATOR left) ready for next calculation. (Note: addition of answer 0 to LSB is automatic and trivial.)
3) If result is positive, multiply by 2, add 1 to the LSB, and store as numerator (in ACCUMULATOR) for next calculation.

If the SUBC instruction is repeated N times, where N is the word length, then the division algorithm is complete. All that remains is to store the lower half of the ACCUMULATOR containing the answer, with the appropriate shift.

There is one minor drawback with the division algorithm—it can only operate with positive numbers. Thus, if negative numbers are to be processed, the division algorithm must work with the absolute values, and restore the correct sign for the quotient in the final answer. A simple technique for achieving this is shown in Example 3.5 together with a complete 16-bit division.

Some of the more recent DSP devices perform the complete division algorithm using a single *multi-cycle* instruction which represents a considerable saving in program code.

TMS EXAMPLE No. 3.5

Example of division, using conditional subtraction.

Begin main program

START

The division procedure using the SUBC instruction requires absolute values. Therefore, the sign of the division result must be calculated in advance

```
ZALS   NUMERA        :load low, no sign extension
XOR    DENOM         :XOR (particularly MSB's)
SACH   SNRSTR,1      :save MSB XOR result
```

If the sign of the two operands is the same, the end result of the division will be positive. If they are different, then the end result will be negative. The above routine causes SNRSTR to hold 1 if the two signs are different, and 0 if they are the same.

```
LAC    DENOM         :load denominator
ABS                  :produce the absolute value
SACL   DENOM         :save it

LAC    NUMERA        :align the numerator
ABS                  :produce the absolute value

RPTK   15            :repeat for 16 bits
SUBC   DENOM         :conditional subtraction
SACL   QUOTNT        :save the absolute result
```

Note that any remainder is located in the high accumulator and can be saved if required

Now restore the sign to the result.

```
LAC    QUOTNT        :load accumulator (1*QUOTNT)
LT     QUOTNT        :set up multiply
MPY    SNRSTR        :multiply with 1 or 0
```

The P register holds (1*QUOTNT) or 0, depending on SNRSTR.

```
SPAC                 :subtract from accumulator
SPAC                 :subtract twice
```

The accumulator holds (QUOTNT) if SNRSTR was 0, and (-QUOTNT) if SNRSTR was 1

```
SACL   RESULT        :save signed result
```

PROGRAM DETAILS:

Instruction cycles	32 for 16 bit division
Program memory words	17 (without initialisation)
Data memory words	5 (all re-usable)

102

Floating-point arithmetic In the vast majority of DSP applications, fixed-point 16-bit arithmetic gives sufficient dynamic range (96 dB) to accommodate most signals without loss of accuracy. There are some cases, however, such as signal squaring which doubles the dynamic range of the signal to be stored, where a fixed 16-bit representation is inadequate. The problem is not that more bits of resolution are required, but that if large and small numbers have to be represented using a fixed format (i.e. fixed decimal place), then, for the small numbers, most of the significant bits have to be zero to signify the relative difference in size between the two numbers.

For example, in Q15 format,

$$23040 = 1011010000000000$$
$$3.5625 = 0000000000000011 \ (= 3)$$

The *quantization error*, as it is called, in the representation of the smaller number can be overcome by using a numerical representation known as floating-point. As the name suggests, this allows the decimal point to be moved, reducing the number of leading zeros. For example, the two numbers shown above can be represented as

$$23040 = 1011010000000000 \times 2^0$$
$$3.5625 = 1110010000000000 \times 2^{-n} \quad (n = 14)$$

Each number (called MANTISSA) has associated with it a number of left shifts (called EXPONENT), in this case 0 and 14, representing the total shift of each number from its original position. Clearly, when a number is stored, the appropriate shift must be stored with it. Rather than use a separate data memory location for this purpose, it is common practice to use only one data word to store both mantissa and exponent. This requires the mantissa to be restricted to say the top 12 bits of the data word, with the exponent stored in the lower 4 bits, e.g.

$$23040 = 101101000000 \qquad 000$$
$$3.5625 = 111001000000 \qquad 1110$$

$$\text{mantissa} \qquad \text{exponent}$$

A number of the more recent processors are specifically designed to handle binary floating-point arithmetic. The earlier devices, including the TMS32020 which are fixed-point processors, require algorithms to convert and process fixed-point numbers in floating-point format. An example of a fixed- to floating-point conversion and multiplication is shown in Example 3.6. The reader is recommended to refer to the operator's manual relating to the device in question to determine the optimum techniques for implementing floating-point arithmetic.

TMS EXAMPLE No. 3.6

An example of converting from fixed to floating point format (normalisation) and multiplying in floating point format.

The result of $a=pi*r^2$ is calculated. Pi is equal to 3.1416 and is stored in floating point format with a mantissa of 0.7854 (=25786 in Q15) and an exponent of 2. The value of r is input and can range between 0 and 0.999... . It is stored in Q15 format

Begin main program

```
START   IN    R,PA0        :input the value of r
        SQRA  R            :square it
        PAC                :result to accumulator
        APAC               :remove extra sign bit

        LARK  AR1,0        :clear normalisation counter
        RPTK  15           :repeat for 16 bits
        NORM               :normalise accumulator

        SACH  R2           :save mantissa
        SAR   AR1,EXPR2    :and exponent
```

Note here that the exponent of r^2 should be a negative value, since it is less than 1. However, it has been stored as a positive value for the time being.

```
        LT    PI           :set up for multiply
        MPY   R2           :do pi*r^2
        PAC                :result to accumulator
        APAC               :remove extra sign bit

        LARK  AR1,0        :clear normalisation count
        NORM               :renormalise if necessary
        SACH  RESULT       :save result mantissa
        SAR   AR1,EXPRES   :save any extra exponent
```

Mantissas are multiplied, exponents must be added (note signs).

```
        LAC   EXPPI        :load exponent of pi
        SUB   EXPR2        :add (negative) exponent of r^2
        SUB   EXPRES       :account for any extra shift
        SACL  EXPRES       :save result exponent
```

Important: When working with floating-point arithmetic, the following rules of multiplication, addition and subtraction apply. To multiply two floating-point numbers, the mantissas are multiplied and the exponents added. For floating-point addition or subtraction, the mantissas must first be shifted to give the same exponent value for both numbers.

3.5.3 Evaluation of f(x)

There are a number of arithmetic functions which cannot reasonably be evaluated by direct binary manipulation, such as $COS(x)$, $LOG(x)$, $\sqrt{x}$, and so on, and alternative approximation techniques have to be adopted. For evaluation of a general function $f(x)$, three main techniques can be applied. The first involves storing a conversion table in data or program memory. The second involves a curve fitting or linear piecewise approximation strategy. The third employs series expansion techniques.

Look-up table method The most straightforward method of evaluating a general function $f(x)$ is to provide in memory a table of $f(x)$ for all values of x likely to be encountered. The data word representing the value x is used to form an address pointer into the data table, and the corresponding value for $f(x)$ read out. Provided sufficient data or program memory is available to give the required accuracy and resolution for $f(x)$, then this approach is the most efficient in terms of algorithm execution time, particularly if the memory addressing scheme for the data table is wisely chosen. Where memory is at a premium, one of the following two techniques may be adopted instead.

Curve fitting Consider the implementation of a digital spectrum analyser using, for example, an FFT algorithm (see Chapter 5). This algorithm generates the frequency spectrum from a number of data points, giving the component magnitudes on a linear scale. In order to convert to a conventional scale in dB, the logarithm of each component magnitude must be calculated. This can be achieved with binary arithmetic by making use of the fact that the function $f(x) = 10 \log(x)$ can be approximated graphically by a series of values along with intermediate slopes and intercepts (Fig 3.1) valid over specific ranges of the input variable x. Thus, to generate the logarithm of a number, the algorithm must firstly determine the range occupied by the input value and then calculate the output value y from the formula for a straight line:

$$y = mx + c$$

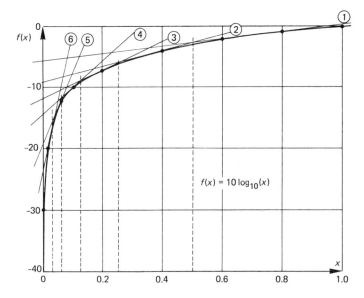

Fig 3.1 Curve-fit for $f(x) = 10 \log_{10}(x)$

where m and c are the predetermined slope and intercept point for the straight line fit over the selected input range. For example, assume an input value of 0.4 for which we require the logarithm (base 10). From Fig 3.1, we note that, for the input range from 0.25 to 0.5, the straight line approximation {2} has a slope of 12.0 and an intercept on the y-axis of -9.0. Making use of the above equation, we can calculate that the logarithm of 0.4 is -4.2 dB. (Actual value is 3.98 dB.) In this example, the ranges into which the input is divided are based on powers of two, i.e. 1 to 0.5, 0.5 to 0.25, 0.25 to 0.125, etc. Some functions, however, are best approximated using a linear division of ranges, i.e. 1 to 0.9, 0.9 to 0.8, 0.8 to 0.7, etc., particularly those with a small dynamic range. Examples 3.7 and 3.8 demonstrate the use of both these techniques with the TMS32020.

In some cases, it may be found that the slope and intercept values for all of the ranges of a curve-fit can be represented by a small number of distinct values plus an appropriate shift. This property allows a considerable saving in memory storage for the curve-fit constants, and may also lead to simplification of the curve-fit routine.

As an example, consider the function $y = \sqrt{x}$. The equations for calculating the slope m and intercept values for each range are

$$\text{Slope} = (y_2 - y_1)/(x_2 - x_1)$$

where (x_1, y_1) are the starting co-ordinates of the range being considered, and (x_2, y_2) the end co-ordinates (start of next range).

$$\text{Intercept} = y_1 - mx_1$$

106

Range	x_1	y_1	x_2	y_2	Slope	Intercept
$\frac{1}{16}$ to $\frac{1}{8}$	$\frac{1}{16}$	$\frac{1}{4}$	$\frac{1}{8}$	$\frac{1}{2\sqrt{2}}$	$4(\sqrt{2}-1)$	$\frac{1}{4}(2-\sqrt{2})$
$\frac{1}{8}$ to $\frac{1}{4}$	$\frac{1}{8}$	$\frac{1}{2\sqrt{2}}$	$\frac{1}{4}$	$\frac{1}{2}$	$2(2-\sqrt{2})$	$\frac{1}{2}(\sqrt{2}-1)$
$\frac{1}{4}$ to $\frac{1}{2}$	$\frac{1}{4}$	$\frac{1}{2}$	$\frac{1}{2}$	$\frac{1}{\sqrt{2}}$	$2(\sqrt{2}-1)$	$\frac{1}{2}(2-\sqrt{2})$
$\frac{1}{2}$ to 1	$\frac{1}{2}$	$\frac{1}{\sqrt{2}}$	1	1	$1(2-\sqrt{2})$	$1(\sqrt{2}-1)$

Table 3.1 Slope and intercept values for $f(x) = \sqrt{x}$ curve fit

Using the two above equations yields the table of slope and intercept values shown in Table 3.1. Note that there are only two distinct values, $(\sqrt{2}-1)$ and $(2-\sqrt{2})$, plus appropriate shifts, to represent the slopes and intercepts for all four ranges.

The curve-fitting technique can be applied to a vast range of arithmetic functions and is relatively efficient both in terms of flexibility and algorithm execution time. To emphasise this point, consider the function

$$y = a/\sqrt{b}$$

This can be performed in two ways. The first method is to use a curve-fitting routine to find the square root of b and then perform the division $a/\sqrt{b}$. The second method is to approximate the function $f(b) = 1/\sqrt{b}$ and then multiply the result by a. In terms of program execution time, the second method is considerably more efficient. The curve-fitting routine will take approximately the same time to evaluate both $\sqrt{b}$ and $1/\sqrt{b}$, whereas the division algorithm, as has already been shown, is substantially more complex than simple multiplication. This technique can be used in a wide variety of DSP applications.

Series expansion techniques An alternative approach to the evaluation of complex arithmetic functions is to use an appropriate series expansion, where one exists, to generate an expression consisting of easily manipulated arithmetic operations such as addition, subtraction and multiplication. For example, consider the function $f(x) = 10 \log(x)$. This can be written as the infinite series

$$10 \log(x) = 4.35[(x-1) - \tfrac{1}{2}(x-1)^2 + \tfrac{1}{3}(x-1)^3 - \tfrac{1}{4}(x-1)^4 + \cdots]$$

where $|x-1| < 1$.

TMS EXAMPLE No. 3.7

An example of curve-fitting, using powers of 2 ranges.

A linear piecewise (y=mx+c) curve-fit approximation to the equation f(x)=10log(x) is used to convert values from a linear to logarithmic (dB) scale. Four ranges are used, covering the linear values between 2^0 and 2^{-4}, which correspond to the logarithmic values between 0 and -12.04 dB. Linear values are in Q15 format, and logarithmic values in Q11 format.

Insert the lookup table of intercept (c) values in program memory. Note that after the linear values have been normalised, the value of curve-fit slope (m) for each range is found to be the same, and thus a lookup table of slopes is not needed.

```
CFTBL     DATA   -12330,-18495,-24460,-30825
```

The above values correspond to intercepts (in Q11 format), calculated for each range.

Begin the main program.

```
START    ZALH   LINEAR        :load in the linear value
         LARK   AR1,0         :clear the normalise counter
         RPTK   2             :do 3 times (gives 4 ranges)
         NORM                 :normalise
```

The normalisation process increases the accumulator in powers of two, up to the maximum range (0.999.. to 0.5) or the end of the repeat count.

```
         SACH   X             :save the scaled value
         SAR    AR1,RANGE     :save the range
```

The normalise counter indicates which powers of two range the original value was in.

```
         LALK   CFTBL         :point to the lookup table
         ADD    RANGE         :add the range offset
         TBLR   C             :transfer the intercept value

         LAC    C,15          :load accumulator with '+c'
```

Note that the intercept value has been scaled to the same range as the following multiplication product.

```
         LT     X             :set up the multiply
         MPY    M             :(m=12330) do 'm*x'
         APAC                 :add to c
         SACH   DB,1          :save the result (y=mx+c)
```

PROGRAM DETAILS:

Instruction cycles	19 for 4 ranges
Program memory words	15 (without initialisation)
	plus 4 for table
Data memory words	6 (5 are re-usable)

TMS EXAMPLE No. 3.8

An example of curve-fitting, using linear ranges.

A linear piecewise (y=mx+c) curve-fit approximation to the equation f(x)=10log(x) is used to convert values from a linear to logarithmic (dB) scale. Four ranges are used, covering the linear values between 0.999.. and 0.2, which correspond to the logarithmic values 0 and -6.9897 dB. Linear values are in Q15 format, and logarithmic values in Q11 format.

Insert the lookup table of slope (m) and intercept (c) values in program memory.

```
CFTBL    DATA    30833,-20483        :linear range 0.2 to 0.399
         DATA    18031,-15362        :0.4 to 0.599
         DATA    12794,-12220        :0.6 to 0.799
         DATA    9924,-9924          :0.8 to 0.999
```

Begin the main program.

```
START    LT      LINEAR       :set up for multiply
         MPYK    5            :5 ranges, 0 to 0.99..
         PAC                  :product to accumulator
         SACH    RANGE,1      :save integer portion
```

The integer portion of the multiplication product gives a value for the curve-fit range, from 4 for the range 0.8 to 0.999, to 1 for the range 0.2 to 0.399.

```
         LALK    CFTBL-2      :point to the lookup table
```

Note that the value of RANGE equal to zero corresponds to the 5th. range, 0 to 0.199, which is below our lookup table, hence we point to CFTBL-2.

```
         ADD     RANGE,1      :+ offset (2 values per range)
         TBLR    M            :transfer slope value
         ADD     ONE          :update source pointer
         TBLR    C            :transfer the intercept value

         LAC     C,15         :set up '+c' in accumulator
```

Note that the intercept value has been scaled to the same range as the following multiplication product.

```
         LT      X            :set up the multiply
         MPY     M            :do 'mx'
         APAC                 :add to c
         SACH    DB,1         :save the result (y=mx+c)
```

PROGRAM DETAILS:

Instruction cycles	19 for 4 ranges
Program memory words	15 (without initialisation)
	plus 8 for table
Data memory words	6 (5 re-usable)

COMMENTS.

      The linear curve-fit could be divided further, into smaller ranges to improve the approximation. This would require a larger lookup table, but in contrast to the powers of two curve-fit, the execution time would remain the same.

By first finding the value of $(x-1)$ the calculation of the first few terms in the expansion is trivial. Taking only the first six terms yields the answer $10 \log(0.4) = -3.95$ dB, which compares favourably with that achieved using the curve-fitting routine of -4.1 dB. Care must be taken when using series expansion techniques to ensure that the expansions adopted exhibit rapid convergence. If this is not the case, the curve-fitting algorithm is usually more efficient for a given numerical accuracy.

3.6 General processing considerations

AC coupling Any practical A/D converter or sample-and-hold amplifier will always introduce a small amount of dc offset into a signal, irrespective of whether the input waveform is ac or dc coupled. For a number of signal processing applications, particularly those involving modulation (see Chapter 6), this dc offset cannot be tolerated. There are two obvious techniques for eliminating the dc offset. One is to subtract an equal and opposite binary value from the incoming data word, making the assumption that the dc offset is fixed. (This has the same effect as an external dc offset control.) The second is to implement an ac coupling filter within the processor. In this latter case, all that is required is a digital filter representation of the familiar RC network shown in Fig 3.2a. As detailed in Chapter 4, such a filter can be realized using the structure shown in Fig 3.2b. The filter cut-off frequency is governed by the constants used in the digital implementation according to the quoted formulas. A suitable algorithm for implementing the ac coupling filter on the TMS32020 processor is given in Example 3.9.

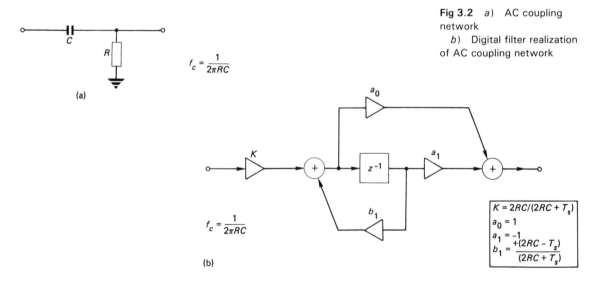

Fig 3.2 a) AC coupling network
b) Digital filter realization of AC coupling network

$$f_c = \frac{1}{2\pi RC}$$

(a)

$$f_c = \frac{1}{2\pi RC}$$

(b)

$$K = 2RC/(2RC + T_s)$$
$$a_0 = 1$$
$$a_1 = -1$$
$$b_1 = \frac{+(2RC - T_s)}{(2RC + T_s)}$$

TMS EXAMPLE No. 3.9

An example of AC-coupling, using a first order filter section.

A high-pass filter with a cutoff frequency of 100 Hertz is implemented, at a sampling rate of 8 kilohertz.

The values of the filter constants are shown below.

$$B1 = 0.9245 \; (30293 \; in \; Q15)$$
$$K1 = 0.0755 \; (2475 \; in \; Q15)$$
$$A1 = -1.0 \; (-32768 \; in \; Q15)$$
$$A0 = +1.0 \; (32767 \; in \; Q15)$$
$$K2 = 12.2389$$

K2 is implemented by a combination of multiplication and left shifting. (Note that 12.2389 is equivalent to 1.5298*8)

Begin the main program.

```
ADC       EQU    PA0              :input through port 0
DAC       EQU    PA1              :output through port 1

DELAY     EQU    149              :count for delay loop


START     IN     INPUT,ADC        :input new signal

          LT     INPUT            :input to filter
          MPYK   K1               :filter constant
          LTP    ACDEL            :accumulate and set multiply
          MPY    B1               :filter constant
          APAC                    :accumulate
          SACH   ACDEL,1          :save new delay result
```

Note that leaving the value of the new delay in the accumulator is equivalent to multiplying it by A0 (= 1.0) and accumulating, so A0 need not be stored.

The T register still holds the value of the old delay.

```
          MPY    A1               :filter constant
          APAC                    :result to accumulator
          SACH   OUTPUT,4         :save and scale result
```

The filter output has been saved with 3 additional left shifts - equivalent to multiplying by 8.

Now complete the output scaling by calculating 0.5298*(8*output) + 1*(8*output).

```
          LT     OUTPUT           :set up multiply
          MPY    K2               :K2 = 0.5298 (17363 in Q15)
          PAC                     :transfer to accumulator
          ADD    OUTPUT,15        :add 1.0*OUTPUT (8*output)
          SACH   OUTPUT,1         :save result

          OUT    OUTPUT,DAC       :output AC-coupled result
```

Implement delay to achieve 8 kHz sampling.

```
          LRLK   AR1,DELAY        :load delay counter
          NOP                     :padding
          NOP
          NOP
```

111

```
PAUSE      NOP                         :pad out delay loop
           NOP
           BANZ   PAUSE                :decrement counter and loop

           B      START                :back round after 625 cycles
```

PROGRAM DETAILS:

Instruction cycles   18 for I/O and filter

Program memory words   16 (without initialisation
                              or delay)

Data memory words   6 words

Overload considerations One of the major sources of error in a
DSP circuit implementation is that of signal overload. The possibility of overflow within a signal processor is discussed in Chapter 1,
where the concepts of sign extension and accumulator saturation are
introduced. With most of the available DSP devices, it is possible to
define a mode of operation where any numerical overflow within the
accumulator (due to the addition of two large numbers for example)
results in the accumulator value saturating at either the maximum
positive or the maximum negative value as appropriate. Operation of
the DSP in this mode is strongly recommended unless the application
requires otherwise. This 'safety net' works perfectly provided the
accumulator contents are stored without further left shifts. If the
latter requirement is not observed, the value to be stored can still
overload as the shifting process takes place outside the accumulator.

An example of the error which can result is shown in Fig 3.3,
where overflow has caused the MSB of the stored sinewave values to
be lost so that a magnitude bit is interpreted as the sign bit. If extra
gain (left shift) is required prior to data storage, it is essential to
ensure that overflow does not occur as a result. This can be achieved
in one of two ways. The magnitude of the waveform can be increased
by repeatedly adding the accumulator value to itself, whereby
overflow protection is automatically realized (Example 3.10).

Alternatively, the accumulator can be forced into saturation by
adding (for positive numbers) or subtracting (for negative numbers)
a predetermined constant value which is subsequently subtracted (or

Fig 3.3 Effect of overflow
on waveform characteristics

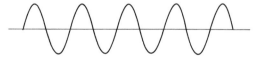

(a) **Desired waveform**   (b) **Waveform with SIGN BIT overflow**

added) to leave a clipped version of the original waveform (Fig 3.4). As the maximum value of the waveform is now well defined, storage using an appropriate left shift will not result in an overflow error. This process is illustrated in Example 3.11.

In some applications, it is desirable to remove the amplitude variations from a signal altogether, and retain only the sign of the

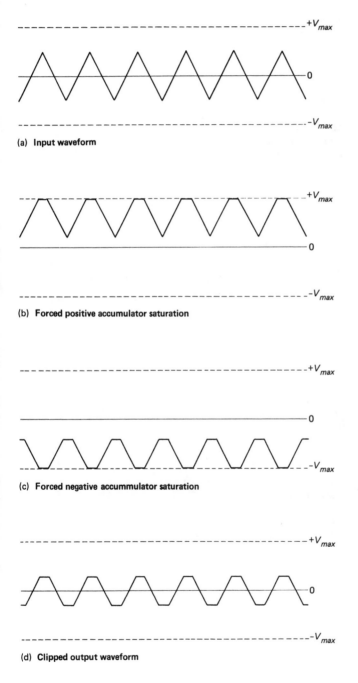

(a) Input waveform

(b) Forced positive accumulator saturation

(c) Forced negative accummulator saturation

(d) Clipped output waveform

Fig 3.4 Waveform clipping by forced accumulator saturation

113

TMS EXAMPLE No. 3.10

An example of protecting against overflow when increasing the value of a signal.

Begin the main program.

START	SOVM		:ensure overflow mode on
	LT	SIGNAL	:set up multiply
	MPY	VALUE	:multiply by some value
	PAC		:transfer to accumulator
	APAC		:remove extra sign bit
	SACH	RESULT	:save result

Leave result in accumulator and add it to itself to increase its value.

	ADDH	RESULT	:add to accumulator
	ADDH	RESULT	:add again (now times 3)
	SACH	RESULT	:save result without shift

TMS EXAMPLE No. 3.11

An example of protecting against overflow when increasing the value of a signal, using a limit mask.

Begin the main program.

START	SOVM		:ensure overflow mode on
	LT	SIGNAL	:set up multiply
	MPY	SCALE	:perform multiply
	PAC		:transfer to accumulator
	APAC		:remove extra sign bit

Add limiting mask to accumulator to allow save with extra left shifting

	ADDH	LIMIT	:add mask (>7800 in this case)
	SUBH	LIMIT	:remove again

Subtract for negative values.

	SUBH	LIMIT	:subtract mask
	ADDH	LIMIT	:add back again

Can now save result with a left shift of 4

| | SACH | RESULT,4 | :save result with extra shift |

TMS EXAMPLE No. 3.12

Two examples of hard limiting.

Begin the main program.

LIMIT	EQU	>FFFE	:define XOR mask
	SSXM		:ensure sign extension on

First method involves sign-testing

START1	LAC	VALUE	:load value into accumulator
	BGEZ	POSITV	:test sign of value

If routine has not branched, value is negative - load accumulator with most negative value

	LALK	>8000	:most negative value
	B	SAVRES	:branch to save result

Value is positive if routine branches here - load accumulator with most positive value.

POSITV	LALK	>7FFF	:most positive value

Save accumulator to give hard limited result

SAVRES	SACL	RESULT	:save hard limited result

Second method involves bit masking.

START2	LAC	VALUE	:load accumulator

Sign extension causes the high accumulator to hold >FFFF for negative values or 0000 for positive values

	XORK	LIMIT,15	:invert 15 lsb's of high acc.
	SACH	RESULT	:save hard limited result

PROGRAM DETAILS:

Instruction cycles	6 or 8 for first method 4 for second method
Program memory words	10 for first method 4 for second (without initialisation)
Data memory words	2 for either (re-usable)

data value; this is commonly termed *hard limiting*, as opposed to *soft limiting* or *clipping* as outlined above. One of the simplest algorithms for hard limiting involves testing the sign bit of the 2's complement word to determine the signal polarity ($0 = +$ve, $1 = -$ve), and placing the appropriate signed value in the output register. An alternative technique exploits the sign extension mode of the TMS320-20 with suitable bit masking. These two techniques are illustrated in Example 3.12. Some more recent processors feature a single LIMIT command in their instruction set.

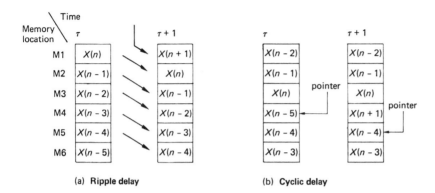

Fig 3.5 Delay mechanisms

(a) **Ripple delay**

(b) **Cyclic delay**

Time delay Digital delay of a waveform is easily achieved by storing successive data samples in memory and reading out the appropriate sample after the required delay. Delay algorithms are greatly simplified if the data samples are stored in adjacent memory locations allowing simple 'ripple' data moves or cyclic memory access techniques to be used for data handling (Fig 3.5). For short delays, typically less than 10–20 sample periods, the ripple delay technique is most efficient in terms of execution time. This algorithm operates on the data exactly like a Bucket Brigade or CCD delay line used for analog signal delay, whereby the oldest memory location is updated with the contents of the second oldest memory location, which in turn is filled with the third oldest value, until all data words in the delay line have been successively shuffled through, and the most recent sample value stored. Clearly this can be a lengthy process for long delays since the operation is sequential rather than parallel in nature.

For the implementation of long delays, the cyclic technique is the more efficient in terms of execution time. In this case an address counter is used to cycle through the memory locations, extracting the data word at a given location corresponding to the required delay, and inserting the most recent sample in its place, i.e. only one fetch and store sequence per program cycle compared with N for the ripple counter method, where N is the number of delay elements used. The

116

address counter is then incremented (or decremented) to point at an adjacent memory location which will be of the correct delay, and the process is repeated. This technique should be adopted for delays in excess of 10–20 sample periods. TMS32020 implementations for both delay algorithms are given in Examples 3.13 and 3.14. If very large delays are required, then external memory has to be used for data storage and account must be taken of any additional read/write delay to off-chip memory when choosing the optimal algorithm. (Note: The ripple and cyclic delay techniques require the same amount of memory for data storage.)

TMS EXAMPLE No. 3.13

An example of imposing time delay on a signal, using a 'ripple' technique.

A signal is input, then output after a delay of 10 sampling periods. The delayed input signals are rippled through 10 consecutive data memory locations.

Begin main program.

PG6	EQU	>300	:start of data page 6
INPUT	EQU	1	:input to delay line
OUTP	EQU	10	:output from delay line
START	OUT	OUTP,PA1	:output delayed signal
	LRLK	AR1,OUTP-1+PG6	:set up indirect addressing
	RPTK	9	:repeat for 9 lowest locations
	DMOV	*-	:do the ripple
	IN	INPUT,PA0	:input new signal to delay
	B	START	:back for more

PROGRAM DETAILS:

Instruction cycles	17 for I/O and 10 sample delay
Program memory words	8 (without initialisation)
Data memory words	10 words

COMMENTS.

The use of indirect addressing is necessary when using the repeat mode. Using direct addressing without the repeat mode would allow a saving of 3 instruction cycles, at the expense of increased program memory size.

TMS EXAMPLE No. 3.14

An example of imposing time delay on a signal, using a 'cyclic' technique.

A signal is input, then output after a delay of 20 sampling periods. The delayed input signals are cycled through a buffer of 20 consecutive data memory locations.

Begin main program.

PG6	EQU	>300	:start of data page 6
BUFSTT	EQU	1	:start address of buffer
BUFSTP	EQU	20	:end address of buffer
START	LAR	AR1,DPNTR	:pointer into buffer
	OUT	*,PA1	:output delayed signal
	IN	*+,PA0	:replace old signal with new

Pointer also incremented.

	LRLK	AR0,BUFSTP+PG6	:set up for compare
	CMPR	2	:check if pointer must wrap
	BBZ	SAVPTR	:branch if not
	LRLK	AR1,BUFSTT+PG6	:otherwise wrap pointer around
SAVPTR	SAR	AR1,DPNTR	:save new pointer
	B	START	:back around

PROGRAM DETAILS:

Instruction cycles	13 or 15 cycles
Program memory words	13 (without initialisation)
Data memory words	21 words

COMMENTS.

The above example would execute in the same number of cycles with larger or smaller delay lines - unlike the previous 'ripple' example.

Counters Binary counters form the basis of many DSP algorithms, ranging from cyclic delay loops to sinewave generators to digital filters. Most DSP devices have dedicated registers, the auxiliary registers, which can function as counters and pointers. In some cases, a specialised REPEAT instruction exists for multiple pipelined execution of single instructions (or in some cases groups of instructions). Both techniques involve the auto incrementing or decrementing of a register until a predefined count value is reached at which point the counter loop is broken or reset.

An alternative counter design which does not rely on auxiliary registers makes use of a standard 16-bit data memory location to hold the count value, updated via the accumulator. The count step size is similarly stored as a second 16-bit word and easily and accurately modified if a variable count rate or cycle time is required (e.g. voltage controlled oscillators, Chapter 6). The count length and start/stop values can be set by loading an initial value into the counter register (count start) and comparing subsequent count values with the value held in a count stop register. When the stop value is reached (or exceeded), the counter loop is broken or the counter register re-initialized and the count is repeated.

To reduce the memory requirement, the stop count value can be set to zero, and use made of the resident zero accumulator branch commands (e.g. BLEZ: Branch if accumulator is less than or equal to zero) to determine count end. Similarly, the counter can be initialized at full-scale value to remove the need for a start register, the appropriate step size being chosen to achieve the desired count rate. This process can be taken a step further for repetitive counters by using a 'modulo' counting procedure to eliminate the stop/reset comparison stage entirely. The algorithm makes use of the fact that when a binary counter overflows and the overflow bits are ignored (or masked off) then *wrap-around* occurs (i.e. a large positive number becomes a large negative number using two's complement arithmetic), and the count is thus automatically repeated. This process is illustrated in Fig 3.6. If the count length does not require

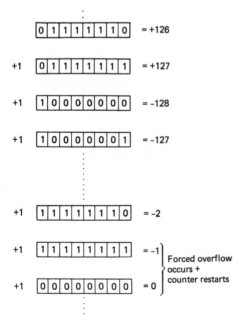

Fig 3.6 Modulo 256 counter (2's complement notation)

TMS EXAMPLE No. 3.15

An example of implementing a modulo 256 counter.

Define modulo mask.

```
MASK        EQU     >00FF                   :mask for ANDing
```

Begin main program.

```
START       LAC     COUNT                   :load current count
            ADD     STEP                    :update it
            ANDK    MASK                    :mask off lower 8 bits
            SACL    COUNT                   :save new count value
```

the full 16 bits of a data word, called a modulo 65536 (2^{16}) counter, then a 'forced' overflow condition can be achieved at counts of less than 16 bits by masking off the unwanted higher-order bits using the appropriate LOGICAL operation. Implementation of a modulo 256 or 8-bit counter on the TMS320-20 is demonstrated in Example 3.15.

3.7   Anti-alias and reconstruction filtering

Before proceeding to the more complex signal processing tasks of filtering, frequency transformation, modulation, etc, outlined in subsequent chapters, the requirements for external signal filtering are discussed. Even for the very simplest of algorithms operating on an analog input waveform and producing an analog output waveform, it is vital to appreciate the effects of signal aliasing on system behaviour. This topic is discussed fully in section 2.6 which highlights the need for the A/D and D/A conversion stages.

3.7.1   The anti-aliasing filter

The theory presented in Chapter 2 demonstrates that distortion-free recovery of a sampled reference signal is only possible when the reference signal is entirely bandlimited to half the sampling rate. Even if the input signal is naturally bandlimited, an anti-aliasing filter is still advisable to reduce out-of-band noise aliasing into the wanted frequency band.

Ideally, the anti-aliasing filter should have a flat amplitude and linear phase response over the bandwidth of the signal, and infinite attenuation at the half sampling rate and beyond. A practical analog filter can clearly not meet this specification, having, amongst other faults, a finite transition region. Although the filter transfer function in the transition range need not affect the wanted signal, any non-zero response beyond the half sampling frequency will permit high-frequency input components and noise to alias into the wanted band.

In the absence of an anti-aliasing filter, a white noise signal with power spectral density N_0 and bandwidth nB_s (where B_s is the sampling rate and n is integer) will result in a recovered signal with noise spectral density of $2nN_0$ occupying a bandwidth $B_s/2$. In other words, the noise power in the recovered, bandlimited signal is the same as in the entire bandwidth of the input signal, assuming no processing of the sampled signal.

Non-ideal passband magnitude and/or phase characteristics in the anti-aliasing filter (and also in the reconstruction filter) can be compensated for to a greater or lesser extent within the DSP. Whilst it is difficult to synthesize a linear phase constant-gain characteristic using active or passive lumped element analog filters, it is relatively easy to compensate for non-linear phase and gain in a discrete process (cf. Chapter 4). One of the most convenient forms of anti-aliasing filter is the switched capacitor type since the filter bandwidth can be varied according to the internal or external filter clock. Typically, the clocking rate is 50 to 100 times the filter cut-off frequency, and anti-aliasing of the switched capacitor filter itself can be achieved with a simple RC network.

A technique for relaxing the specification of anti-aliasing filters, known as *oversampling*, can be adopted if sufficient processing time is available. This technique is outlined in section 4.7.

Summarising, the sample rate must be at least twice the signal bandwidth, but there may be advantages in oversampling to ease the specification for the anti-aliasing filter.

3.7.2 The reconstruction filter

A continuous signal can be recovered without distortion from its ideally sampled version by low-pass filtering. The ideal reconstruction filter has a flat gain response and linear phase characteristic in the passband extending from dc to the half sampling rate and infinite attenuation in the stopband beyond. As with anti-aliasing filters the task of synthesizing the reconstruction filter is made easier if the sampling rate is more than double the signal bandwidth, since a finite transition region can then be accommodated. Once again switched capacitor filters are to be preferred because of their programmable cut-off frequency and physical compactness. Any departure of the filter characteristic from the ideal will introduce spectral distortion.

The signal reconstruction process is complicated by the fact that we do not have direct access to an ideally sampled output signal (zero-width weighted impulses). Instead the D/A converter outputs constant-width pulses whose amplitudes correspond to the weights of the embedded ideally sampled signal. In other words, before we have been able to get access to the output it has already been passed through a filter with impulse response:

$$h(t) = \text{rect}[(t - T_w/2)/T_w]$$

where T_w is the pulse width (cf. section 2.3). This is equivalent to a filter with transfer function:

$$H(f) = T_w \cdot \text{sinc } [fT_w]$$

Such a filter is called a *zero order hold*. Table 3.2 gives the excess attenuation at the half sampling frequency caused by the zero order hold as a function of the normalized pulse width T_w/T_s. The smaller the pulse width the less the unwanted shaping, but also the smaller the output signal amplitude. For most D/A configurations, the normalized pulse width is unity.

T_w/T_s	Excess attenuation at $f_s/2$ (dB)	DC gain compared with $T_w/T_s = 1$ (dB)
1	∞	0
0.5	− 0.9	− 6
0.33	− 0.4	− 9.5
0.25	− 0.22	− 12
0.1	− 0.03	− 20

Table 3.2 Excess attenuation at $f_s/2$ due to zero order hold of width T_w

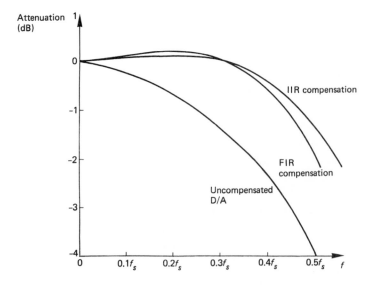

Fig 3.7 D/A equalization filter response.

To counteract the filtering effect of the D/A, one solution is to build into the discrete processor a compensating filter which boosts the response between $f_s/3$ and $f_s/2$ to partially compensate for the roll-off caused by the zero order hold. Two suitable transfer functions (FIR and IIR) are

$$H(z) = -(1/16) + (9/8)z^{-1} - (1/16)z^{-2}$$

and

$$H(z) = \frac{(9/8)}{1 + (1/8)\ z^{-1}}$$

Fig 3.7 summarises the performance of the filters. The compensation can also be built into the switched capacitor reconstruction filter itself.

4 Digital filters

4.1 Introduction

Signal filtering in one form or another is probably the most common application of DSP. Not only does DSP enable the design of filters with consistently reproducible and well defined characteristics, but the ease and flexibility of complex filter implementation is unparalleled. This latter feature is due, at least in part, to the wealth of computer-based filter design software now available. Filter specifications which are particularly difficult or even impossible to achieve using analog techniques, such as linear phase characteristics, Hilbert transform filters (wideband $90°$ phase shifts) and differentiators, can be routinely realized by discrete means.

A major limitation of the discrete approach, however, is that the filter transfer function is periodic in the frequency domain. This means that *a digital filter can only ever be an approximation of the equivalent analog filter*, with the best match occurring at frequencies well below the half sampling rate.

The first two sections of this chapter are intended to provide the reader with sufficient basic knowledge of digital filter design to be able to effectively use one of the many CAD packages on the market. (Fortunately this is a relatively painless process.) The next two sections of the chapter provide a systematic overview of digital filter design techniques to enable the reader to write his or her own support software for particular specialist filter design tasks.

Some familiarity with the contents of Chapter 2 is assumed, particularly the distinction between continuous and discrete transfer functions, the relationship between transfer function and impulse response, and the method of determining frequency and phase responses from the s-plane or z-plane pole/zero plots as appropriate.

Finally, the effect of fixed point, finite word length representations of the signal samples and filter coefficients is discussed. Various design strategies are proposed for handling coefficient quantization and intermediate result round-off; both an unavoidable consequence of using fixed-point arithmetic. This section can be omitted if a floating-point device is to be used.

4.2 Getting started with simple filters

CAD packages are used extensively for the design of digital filters and it is quite possible to design very sophisticated filters knowing only the rudiments of the supporting theory. This section of the chapter provides just enough of that theory to enable the intelligent use of a suitable filter design software package. The treatment is centred around simple first- and second-order filters—useful in their own right—which can be readily designed with just a calculator. They are in fact the discrete equivalents of simple RC and RLC analog designs.

4.2.1 The RC low-pass filter and its digital (discrete) equivalent

Analog filters can be synthesised from R, L and C circuit elements. A very simple example is the RC low-pass filter shown in Fig 4.1, having the transfer function

$$H(s) = 1/(1 + RCs)$$

and impulse response

$$h(t) = (1/RC)\exp(-t/RC)$$

The filter has a single pole at $s = -1/RC$. Both magnitude and phase response can be deduced from the s-plane pole plot using the method outlined in section 2.7.1. The following sections examine a number of techniques for realizing this RC filter response using digital signal processing means.

4.2.2 An intuitive approach to digital filter design

The operation of a low-pass filter is basically that of waveform averaging over a time interval, the averaging period increasing as the filter cut-off frequency is reduced. A discrete approximation to an RC low-pass filter can thus be realized by an equivalent averaging process. The type of averaging is dictated by the properties of the analog filter. We know from experience that the RC filter has only a short-term memory, i.e. once the input has been removed, the output dies away exponentially. The discrete averaging process should therefore mimic this type of response.

A simple discrete sample averager can be implemented using the feedback structure shown in Fig 4.2a. The most recent input sample value is summed with a weighted previous sample held in memory, and the result used to update the memory contents. If the weighting factor b_1 is set to zero, then the output is simply a delayed version of the input, and no filtering/averaging is performed. If b_1 is greater

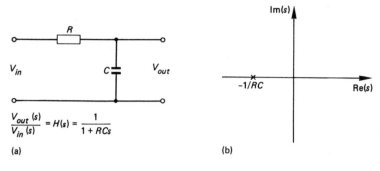

Fig 4.1 *a*) An RC low-pass filter
  b) *s*-plane plot
  c) Magnitude response
  d) Phase response

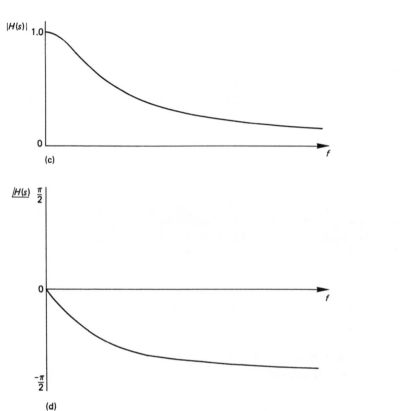

than zero, then the output is conditioned by the previous stored value, and some form of averaging occurs. The greater the weighting factor, the greater the influence of past sample history on the system response. In other words, if the input sample values are varying slowly, then they will eventually have some effect on the stored value, and will be mimicked at the output. If, on the other hand, the input samples vary rapidly in succession, the output cannot directly respond and a filtering action results. The structure can thus be seen to function as a low-pass filter, with response conditioned by the

(a)

(b)

(c)

Fig 4.2 *a*) Basic averaging circuit realization
b) Gain compensated design
c) Complementary high-pass filter realization

weighting factor b_1. This structure is sometimes termed a *leaky accumulator*. When b_1 exceeds unity, the feedback structure becomes unstable. (If b_1 is less than zero, then a high-pass filtering action results, becoming unstable as b_1 exceeds -1.)

If we excite the circuit with a single 'unit pulse', it is fairly clear that the output from the 'filter' will gradually die away, but never reach zero. In other words, the pulse response is infinite. This gives rise to the term *infinite impulse response* filter (IIR).

The steady state gain of the system for $b_1 > 0$ will be maximum when all the input samples are of constant value, and is given by

$$G_{max} = 1/(1 - b_1) \qquad b_1 \geqslant 0$$

Thus, to achieve unity gain, as for the analog *RC* filter, the input must be scaled by a factor $k = 1 - b_1$ (Fig 4.2*b*). Note that the scaling is applied to the input rather than the output in order to prevent system overload.

This intuitive approach to filter design has in fact led us to the *impulse response equivalent* of the analog *RC* filter which will be described more fully in the next section. This structure is also known as an *exponential averager* (cf. section 6.7).

One further feature is that the equivalent high-pass *CR* filter response can be approximated by simply subtracting the low-pass filter output samples from the input samples as shown in Fig 4.2*c*.

4.2.3 Design by impulse response equivalence

A DSP system, as depicted in Fig 2.6, can be used to synthesize an end-to-end *continuous system transfer function*, which we designate $H_c(s)$. At the heart of the DSP system is the discrete process, which can be described by its *discrete transfer function $H_d(z)$*.

Suppose it is desired to have $H_c(s)$, as produced by the DSP system, match as closely as possible the ideal transfer function $H(s)$ of the analog *RC* filter. The first point to note is that $H_c(s)$ can never be identical with $H(s)$. This is because the DSP system can only distinguish frequencies up to half the sampling rate, whereas the analog filter response extends to infinity. Matching $H_c(s)$ to $H(s)$ must therefore always involve a compromise. Either the discrete filter order can be fixed (typically to be the same as the order of $H(s)$), in which case the user must accept the consequent sacrifice in performance, or the accuracy of the match can be fixed, in which case the resulting level of filter complexity must be accepted. In the next few subsections some alternative compromise strategies will be explored with reference to the *RC* low-pass filter.

We begin by attempting to use a first-order (one delay element) discrete system to match the behaviour of the *RC* low-pass filter (which is also first order). Fig 4.3 gives the model for the most general first-order linear discrete process. The convention has been adopted of labelling the feedforward and feedback coefficients (the a's and b's respectively) with a subscript identifying the associated number of unit delays. When filters consisting of cascaded first- and second-order sections are considered in subsequent discussion, a second subscript will be used to identify the stage.

(Care should be taken when using books and design packages as a number of different labelling schemes are in common usage.)

For the system shown in Fig 4.3, it is easy to see that if we make

$$a_0 = 1/RC \qquad a_1 = 0 \qquad b_1 = \exp(-T_s/RC)$$

and $\quad k_1 = k_2 = 1.0$

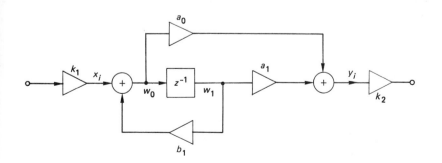

Fig 4.3 Universal first-order filter section

129

then the 'unit pulse response' of the process (cf. section 2.3) is

$$h_d(n) = (1/RC)\exp(-nT_s/RC) \qquad n \geq 0$$
$$= h(nT_s)$$

where T_s is the sampling rate.

Comparison of this result with the analog filter impulse response, $h(t)$, shows that the discrete system unit pulse response consists of a sequence of numbers which are identical with samples of $h(t)$, taken at T_s second intervals. Filters designed with this characteristic are termed *impulse response invariant*. Unfortunately, this property does not result in the end-to-end equivalent continuous system response $h_c(t)$ being the same as $h(t)$. It is in fact the convolution of the samples $h_d(n)$ with $\text{sinc}((t - nT_s)/T_s)$. The desired and actual impulse responses will be equal at the sampling moments, but not in

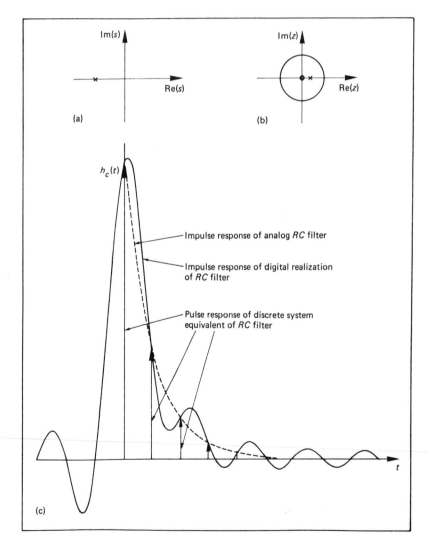

Fig 4.4 First-order digital filter design by impulse response invariance:

a) Desired *s*-plane plot $H(s)$

b) Equivalent *z*-plane plot $H_d(z)$

c) Actual impulse response of equivalent analog system $h_c(t)$

Im(s)

Re(s)

(a)

Im(z)

Re(z)

(b)

$h_c(t)$

Impulse response of analog *RC* filter

Impulse response of digital realization of *RC* filter

Pulse response of discrete system equivalent of *RC* filter

t

(c)

between (Fig 4.4). The part of the specification that is sacrificed in this design is thus the impulse response match in between the sampling moments.

The frequency response of the RC equivalent filter can be determined from the system impulse response (section 2.5) and is given by

$$H_c(2\pi f) = (1/T_s) \cdot H(2\pi f) \star \sum_{k=-\infty}^{\infty} \delta(f - kf_s)$$

As shown in Fig 4.5 the frequency response is periodic and deviates from the ideal RC filter response due to aliasing (with a scaling factor of $1/T_s$). The scaling factor (which can be very small for typical sampling rates) is normally compensated for by multiplying all the feedforward coefficients (the a's in Fig 4.3) by T_s. This is in any case desirable in fixed-point implementations using Q15 format to achieve coefficients near unity.

The discrete transfer function $H_d(z)$ that corresponds to the system in Fig 4.3 can be found by writing the difference equation, and taking its z-transform. We have

$$w_0(n) = x(n) + b_1 \cdot w_0(n-1)$$
$$w_1(n) = w_0(n-1)$$
$$y(n) = a_0 \cdot w_0(n) + a_1 \cdot w_1(n)$$

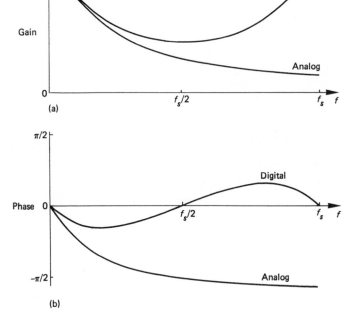

(a)

(b)

Fig 4.5 A first-order analog low-pass filter response and the response of its digital equivalent designed by impulse response invariance:
 a) Magnitude characteristic
 b) Phase characteristic

Hence

$$H_d(z) = Y(z)/X(z) = [a_0 + a_1 \cdot z^{-1}]/[1 - b_1 \cdot z^{-1}]$$
$$= [a_0 \cdot z + a_1]/[z - b_1]$$

In order to match the RC filter response, and taking account of the gain correction discussed above, we have

$$H_d(z) = (T_s/RC) \cdot z/[z - \exp(-T_s/RC)]$$

The associated z-plane plot is shown in Fig 4.4a from which the frequency and phase response (Figs 4.5a and 4.5b) could have been found graphically using the procedure discussed in section 2.7.3, or by making the substitution

$$H_c(2\pi f) = H_d(z)|_{z = \exp(j2\pi fT_s)}$$

The discrete transfer function consists of a zero at 0, which has no affect on the magnitude response, but does affect the phase, and a pole at $z = \exp(-T_s/RC)$. As the sampling rate is increased relative to the time constant, the pole gets closer and closer to the unit circle. This is the discrete equivalent of the proximity of analog filter poles to the imaginary axis of the s-plane. In both cases, crossing the axis results in instability. The closer the poles are to the imaginary axis or unit circle the more sensitive the design is to imperfections, such as component tolerance in analog filters and coefficient quantization in digital filters. This simple example brings out the very important point that the sampling rate ideally should be chosen to match the filtering task. When this is not possible (the sampling rate is usually fixed by external considerations), it is sometimes worthwhile to digitally re-sample prior to filtering. This process, indeed the whole question of the sensitivity of the design to imperfections such as coefficient quantization, is considered in sections 4.6 and 4.7.

The dc gain of $H_c(s)$, found by evaluating $H_d(z)$ at $z = 1$, is

$$G_{max} = (T_s/RC)/[1 - \exp(-T_s/RC)]$$

compared with unity for $H(s)$. Again the discrepancy is due to aliasing and reduces as the sampling rate increases. To avoid overflow, it is sensible to reduce the level of the filter input samples by the inverse of the gain factor, $1/G_{max}$. The final filter structure is shown in Fig 4.3 and corresponds exactly to the intuitive filter structure of Fig 4.2b when the forward tap gain a_0 is incorporated with the input scaling factor k.

Apart from its use as a simple low-pass filter, the $H_d(z)$ just described may be used as a leaky accumulator or exponential averager as outlined in section 4.2.2. Note that an ideal integrator is impossible to realize in practice since it would accumulate a dc signal forever, ultimately resulting in overflow. In the exponential averager this problem is avoided by progressively reducing the weight given to the oldest samples.

The impulse response invariant method can be applied to transfer functions of any order provided that the number of poles exceeds the number of zeros by at least one. This constraint restricts the analog prototypes to those whose impulse responses themselves contain no impulse components and immediately excludes all high-pass and bandstop transfer functions. A means whereby a low-pass filter design can be transformed into a high-pass, bandpass or even bandstop design is presented in section 4.2.8.

The properties of the first-order impulse response invariant filter design are summarised in Table 4.1 and Appendix 2. The code for implementing a general first-order filter section using the TMS32020 is given in Example 4.1.

4.2.4 Design by frequency response matching

Ensuring the correct impulse response at the sampling moments by sacrificing the frequency response to aliasing is often not an acceptable compromise. A second, commonly used design approach preserves the general shape of the analog transfer function $H(s)$ by 'pre-warping' the frequency scale. For example if $H(s)$ has a

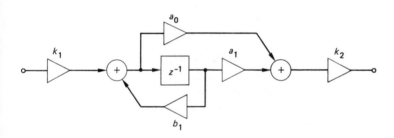

	Low-pass	High-pass (ac coupling)
$H_d(z)$	$\dfrac{kz}{z - b_1}$	$\dfrac{k(z - 1)}{z - b_1}$
a_0	1	1
a_1	0	-1
b_1	$\exp(-T_s/RC)$	$\exp(-T_s/RC)$
k_1	$1 - b_1$	$1 - b_1$
k_2	1	$(1 + b_1)/2(1 - b_1)$
G_{max}	1.0	1.0
	$f_c = 1/2\pi RC$	

Table 4.1 First-order filter design parameters (impulse invariant)

TMS EXAMPLE No. 4.1

General first order filter section.

Begin main program.

```
START      IN       INPUT,ADC        :input signal

           LT       INPUT            :set up multiply
           MPY      K1               :scaling constant
           LTP      DELAY            :accumulate, delay to T
           MPY      B1               :feedback constant
           APAC                      :accumulate
           SACH     DELAY,1          :update delay
```

Note that the T register still holds the old delay value.
The value of A0 is 1.0, hence leaving the accumulator intact is equivalent to multiplying by A0 and accumulating.

```
           MPY      A1               :feedforward constant
           APAC                      :accumulate

           SACH     RESULT,1         :save result
```

The final shift used to save the filter output is dependant on the value of the output scaling factor K2. Further multiplying and shifting may also be required at this point to implement K2 scaling.

```
           B        START            :back around
```

magnitude response that increases from a value of 1.0 at zero frequency to a peak value of 10.0, and after that gradually diminishes to zero at infinite frequency, then, after warping, the magnitude of the equivalent digital filter transfer function $H_c(s)$ will increase from 1.0 at zero to a peak of 10.0 at some intermediate frequency, different from the peak frequency in $H(s)$, and drop to zero at $f_s/2$. This property is illustrated in Fig 4.6.

The warping is achieved within the digital filter equivalent by making the substitution

$$s = \frac{2(z-1)}{T_s(z+1)}$$

Thus, to realize a frequency-warped version of the low-pass RC filter response, the discrete transfer function derived from

$$H(s) = 1/(1 + RCs)$$

would be

$$H_d(z) = \frac{(z+1)/(2RC/T_s + 1)}{z - (2RC/T_s - 1)/(2RC/T_s + 1)}$$

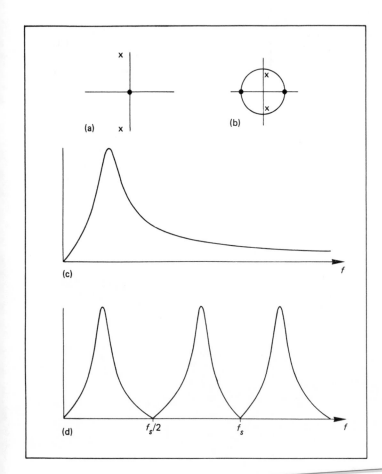

Fig 4.6 Example of frequency axis compression when an analog filter design is converted into a digital filter using the bilinear transform. The entire analog response is compressed into the range $0 < f < f_s/2$:
a),b) s-plane and z-plane plots
c) Analog magnitude characteristic
d) Digital magnitude characteristic

In contrast to the impulse response invariant realization, the gain is unity at dc, falling to zero at half the sampling frequency (warped from zero at infinity for the analog case). The actual shape of the response is a direct consequence of the one-to-one mapping of the entire s-plane imaginary axis onto the z-plane unit circle.

This design approach is called the **bilinear transform method**. Its main weakness is that the 3 dB frequencies of $H_c(s)$ and $H(s)$ are different, related by the expression (cf. section 4.4.1)

$$f_{c\,3dB} = \arctan(\pi f_{3dB} T_s)/\pi T_s$$

A technique known as *pre-distortion* can be used to realize an $H_c(s)$ with the 3 dB frequency in the correct location, by basing the design on an analog prototype with a corner frequency at

$$f_{3dB} = 1/(2\pi\tau) = \tan(T_s/2RC)/\pi T_s$$

The bilinear transform then warps the 3 dB frequency of $H_c(s)$ to $1/2\pi RC$.

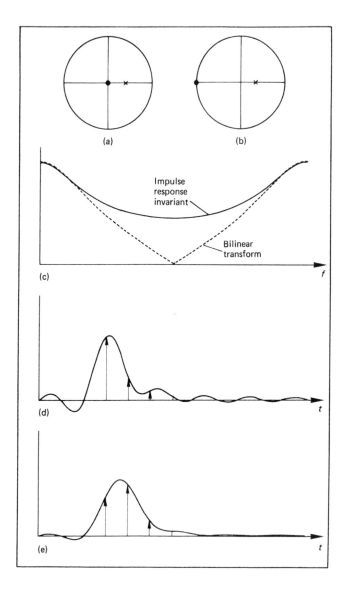

Fig 4.7 Comparison
between first-order bilinear
transform and impulse
response invariant designs:
   a) *z*-plane plot—impulse
invariant
   b) *z*-plane plot—bilinear
transform
   c) Gain response
   d) Impulse response—
impulse invariant
   e) Impulse response—
bilinear transform

Figs 4.7 and 4.8 compare the characteristics of $H_c(s)$ for the two filter types in the time and frequency domains.

The properties of the bilinear transform RC filter realizations are summarised in Table 4.2 and in Appendix 2.

4.2.5 A linear phase version of the RC low-pass filter

In Fig 4.9 an alternative method for generating a sampled version of an arbitrary impulse response is shown. The discrete transfer function $H_d(z)$ has no poles and all the feedback coefficients in the realization are zero (it is non-recursive). The absence of poles means

136

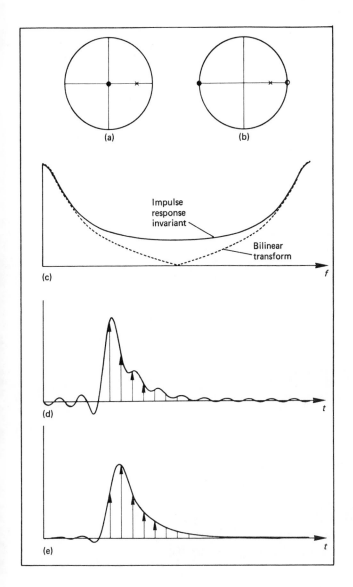

Fig 4.8 As for Fig 4.7 but with double the sampling rate

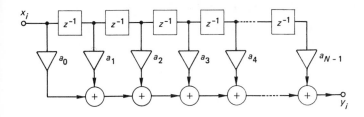

Fig 4.9 All-zero or finite impulse response (FIR) digital filter structure

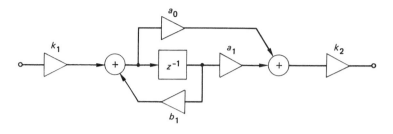

	Low-pass	High-pass (ac coupling)
$H_d(z)$	$\dfrac{k(z + 1)}{z - b_1}$	$\dfrac{k(z - 1)}{z - b_1}$
a_0	1	1
a_1	1	-1
b_1	$(1 - T_s/2RC)/(1 + T_s/2RC)$	$(1 - T_s/2RC)/(1 + T_s/2RC)$
k_1	$(1 - b_1)/2$	$(1 - b_1)/2$
k_2	1	$(1 + b_1)/(1 - b_1)$
G_{max}	1.0	1.0
	$f_c = 1/2\pi RC$	

Table 4.2 First-order filter design parameters (bilinear transform)

that the filter impulse response is of finite duration. The terms **finite impulse response** (FIR) and **infinite impulse response** (IIR) are used to distinguish filter types. Non-recursive filters have all-zero transfer functions and are FIR. On the other hand, recursive filters of N stages have discrete transfer functions with N poles and up to N zeros and are usually, but not always, IIR. In this book FIR will be used synonymously with non-recursive filters and IIR with recursive filters.

There would be little point in using an FIR filter to realize the sampled version of the exponential decay impulse response associated with an RC low-pass filter. We have already seen that a recursive filter can do the job perfectly with only one delay stage compared with the N delays (and multiplications) that would be needed for the FIR filter, and even then the impulse response would be truncated at $t = NT_s$.

FIR filters come into their own when it is desired to realize an impulse response that cannot be readily associated with a realizable RLC filter. A particular example is the generation of linear phase filters, where the phase response is independent of the magnitude characteristic. This facility is particularly important in the field of digital communications. Fig 4.10 shows how non-linear phase and

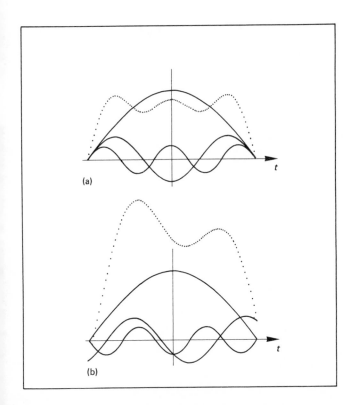

Fig 4.10 Illustration of waveform distortion caused by non-linear phase response of filters:
   a) First three components in square wave synthesis and resultant waveform—linear phase
   b) As for a) but with non-linear phase

hence differential group delay can cause distortion in waveforms passed through conventional analog low-pass filters, even though most of the spectral energy is in the pass band. Linear phase filtering is thus necessary when the time domain waveform distortion must be minimised as in data transmission systems, and also when signals which have been filtered in different ways are later to be compared on an equal delay basis.

Every linear phase transfer function must have an impulse response with *even symmetry* about its mid-point. If the mid-point occurs at $t = T_d$, the slope of the phase characteristic (the group delay) is $-2\pi T_d$ radians/Hz. This property is easily derived using the Fourier transform shift theorem (section 2.4.3).

Returning to the realization of a digital filter which approximates the magnitude characteristic of an *RC* filter but has a linear phase characteristic, we know that the impulse response must be symmetric about some time T_d. A starting point is therefore to sample a symmetric version of the exponential decay for the analog filter as shown in Fig 4.11, i.e.

$$h(t) = 1/\tau \cdot \exp(-|t - T_d|/\tau)$$

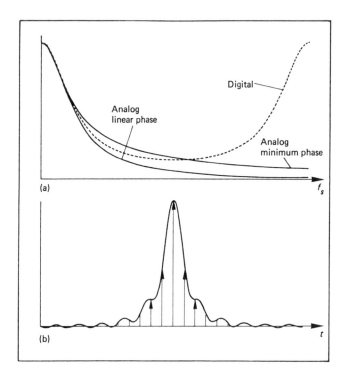

Fig 4.11 Linear phase first-order low-pass filter responses:
  a) Magnitude response — analog and digital
  b) Impulse response of digital realization

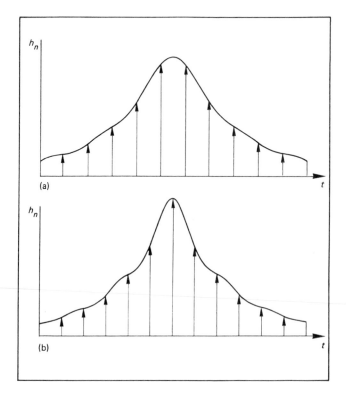

Fig 4.12 Feedforward coefficients of an N-tap FIR filter with a linear phase characteristic:
  a) N even
  b) N odd

Taking the Fourier transform of $h(t)$ we get

$$H(f) = [1/(1 + j2\pi f\tau) + 1/(1 - j2\pi f\tau)] \cdot \exp(-j2\pi T_d/\tau)$$

$$= \frac{2\exp(-j2\pi T_d/\tau)}{1 + (2\pi f\tau)^2}$$

This characteristic has the required low-pass properties and a 3 dB point at $f = \sqrt{(\sqrt{2} - 1)}/2\pi\tau$. Thus, to realize a linear phase version of an analog RC low-pass filter, and FIR filter is needed which has feedforward coefficients which are samples of

$$h(t) = 1/\tau \cdot \exp(-|t - T_d|/\tau)$$

where τ is $\sqrt{(\sqrt{2} - 1)}RC$. If N delay stages are used, then in order to ensure that the filter is causal (i.e. does not have an impulse response before time $t = 0$), a time shift T_d must be introduced where

$$T_d = (N - 1)T_s/2$$

Fig 4.12 shows typical feedforward coefficients or tap weights for even and odd filters. When N is odd there is a sample at the point of symmetry and the associated filter delay is an even number of samples, whereas when N is even there are samples at $T_s/2$ either side of the point of symmetry and the waveform and the total delay is an odd number of half sample delays. The phase characteristics of digital filters whose unit pulse responses have even symmetry about $t = 0$ (unrealizable) and $t = T_d$ (realizable) are shown in Fig 4.13. The equivalent system impulse response $h_c(t)$ is also shown, assuming ideal reconstruction filters.

Note that the order of an FIR filter, that is the number of zeros, is one less than the number of taps.

If the impulse response is made to exhibit an odd symmetry, as shown in Fig 4.14, then the resulting continuous system impulse response will have exactly $\pi/2$ radians phase shift for positive frequencies and $-\pi/2$ for negative frequencies in addition to a linear phase variation. Sampled versions of these impulse responses give discrete filters with a phase characteristic that, apart from the linear variation, alternates between $\pi/2$ and $-\pi/2$ in blocks with frequency span $f_s/2$. This type of filter is used for realizing differentiators and Hilbert transform filters, as will be discussed in section 4.5.6.

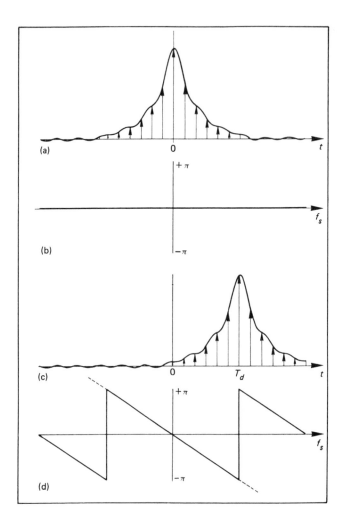

(a)

(b)

(c)

(d)

Fig 4.13 *a*) Unit pulse response of a digital filter with even symmetry about $t = 0$. The continuous curve is the equivalent analog impulse response $h_c(t)$

b) Phase characteristic of filter in *a*)

c) As for *a*) but made causal by delaying response by T_d

d) Phase characteristic of filter in *c*)

4.2.6 Second-order RLC filter types

In analog circuit design, *RLC* components are commonly used to provide simple bandpass and bandstop (notch) functions. For example the transfer function of the circuit in Fig 4.15 is the bandpass function:

$$H(s) = \frac{(R/L)s}{s^2 + (R/L)s + (1/LC)}$$

This transfer function is normally expressed in terms of the resonant frequency in radians per second, $\omega_r = \sqrt{(1/LC)}$, and the Q function, which is the resonant frequency divided by the 3 dB bandwidth, to give

$$H(s) = \frac{(\omega_r/Q)s}{s^2 + (\omega_r/Q)s + \omega_r^2}$$

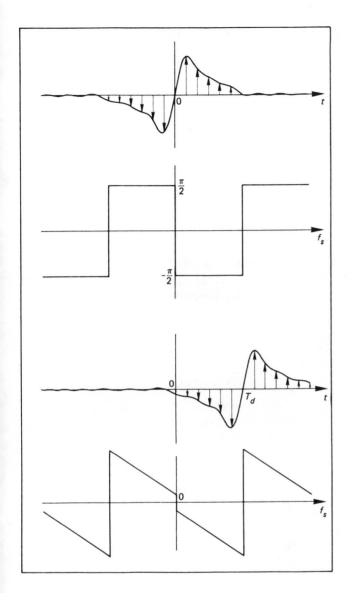

Fig 4.14 As for Fig 4.13 but with a unit pulse response exhibiting odd symmetry about $t = 0$ and $t = T_d$

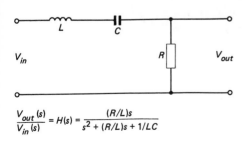

Fig 4.15 RLC unity gain bandpass resonator

$$\frac{V_{out}(s)}{V_{in}(s)} = H(s) = \frac{(R/L)s}{s^2 + (R/L)s + 1/LC}$$

The s-plane plot and the associated magnitude response of the filter are given in Fig 4.16. The exact pole locations are

$$- \omega_r/2Q \pm j\omega_r\sqrt{(1 - 1/4Q^2)}$$

For high Q poles, a reasonable choice for design is the impulse response invariance method. (The useful frequency response is confined to a narrow band, so the aliasing error will be small.) The impulse response of the analog filter can be found by partial fraction expansion and is

$$h(t) = 2\alpha \cdot \exp(-\alpha t) \cdot \cos(\omega_d t) - (\alpha/\omega_d) \cdot \exp(-\alpha t) \cdot \sin(\omega_d t)$$
$$\alpha = \omega_r/2Q \qquad \omega_d = \omega_r\sqrt{(1 - 1/4Q^2)}$$

where α and ω_d are the magnitudes of the real and imaginary parts of the poles. Sampling this impulse response, and taking the z-transform gives us the required discrete system transfer function of the RLC bandpass filter:

$$H_d(z) = \frac{2\alpha \cdot z^2 - 2\alpha \cdot \exp(-\alpha T_s) \cdot \cos(\omega_d T_s) \cdot z}{z^2 - 2\exp(-\alpha T_s) \cdot \cos(\omega_d T_s) \cdot z + \exp(-2\alpha T_s)}$$

The poles have a magnitude of $\exp(-\alpha T_s)$ and occur at angles $\pm \omega_d T_s$ radians; thus the response peaks at ω_d, as in the analog case. The poles can never reach the unit circle, as we would expect since the design was based on the inherently stable (passive) analog filter. The actual distance of the poles from the unit circle is

$$1 - \exp(\omega_r T_s/2Q)$$

which decreases as Q increases and as the resonant frequency is reduced in relation to the sampling frequency.

Note that whereas the analog transfer function $H(s)$ has a zero at 0 Hz, the zero of $H_d(z)$ is not quite at $z = 1$, resulting in a marginal gain at dc. There is also a finite gain at $f_s/2$.

A more useful discrete resonator can be obtained by repositioning the zeros of the transfer function to lie exactly at $z = 1$ and $z = -1$ without changing the pole positions. In the region of the resonance, the response will be little changed (the lengths of the vectors to the new zero locations are not very different), but it will now have the desired property of zero gain at dc and $f_s/2$. The new transfer function, after adjusting for unity gain at the peak response $z = \cos(\omega_d T_s) + j\sin(\omega_d T_s)$, is

$$H_d(z) = \frac{(1 - \exp(-2\alpha T_s))(z^2 - 1)/2}{z^2 - 2\exp(-\alpha T_s) \cdot \cos(\omega_d T_s) \cdot z + \exp(-2\alpha T_s)}$$

To realize this transfer function, an extension of the basic filter

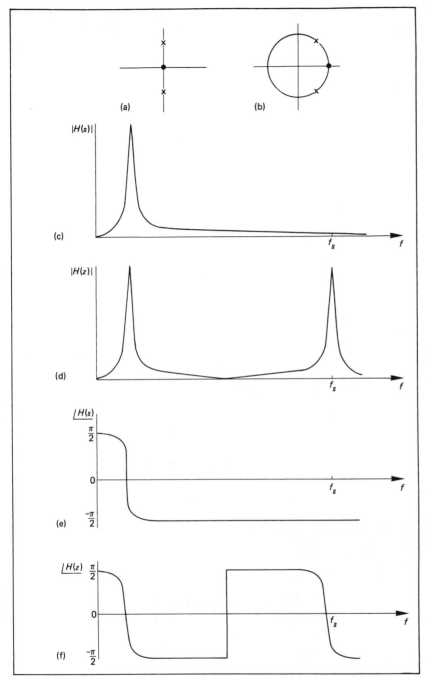

Fig 4.16 Unity gain band-pass resonator responses:
 a) *s*-plane plot of continuous resonator
 b) *z*-plane plot of discrete resonator
 c) Magnitude response of continuous filter
 d) Magnitude response of discrete filter
 e) Phase response of continuous filter
 f) Phase response of discrete filter

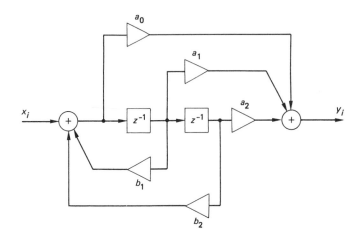

Fig 4.17 Universal second-order filter section

structure is needed as shown in Fig 4.17. Two delay sections are required to realize a second-order discrete transfer function. It is easily shown that the transfer function of this structure is

$$H_d(z) = \frac{a_0 z^2 + a_1 z + a_2}{z^2 - b_1 z - b_2}$$

Note that the feedback coefficients have the opposite polarity to the coefficients of the denominator polynomial of $H_d(z)$—it is easy to make a mistake! The significance of b_2 is that its magnitude is the square of the z-plane pole radius, and b_1 is twice the real part of the pole radius. Appendix 1 and Table 4.3 give the feedforward and feedback coefficients needed (as derived by comparing the above two transfer functions) for implementing the discrete equivalent of the unity gain bandpass resonator in terms of the desired Q and resonant frequency. Also documented are the coefficients for a second-order notch and second-order allpass filter section, the latter playing an important role in phase equalization networks.

A DSP realization of the second-order filter section is given in Example 4.2.

4.2.7 Design by direct pole and zero placement

When it is necessary to realize a filter characteristic with specific features, suitable locations for the poles and zeros of $H_d(z)$ can often be deduced directly. This has already been demonstrated in the previous examples where infinite attenuation at a frequency was achieved by placing unit circle zeros at the appropriate angles. In this section two more examples of this technique are given.

The previous bandpass filter design can easily be adapted to form a notch by placing a pair of zeros on the unit circle at the angle $\omega_d T_s$.

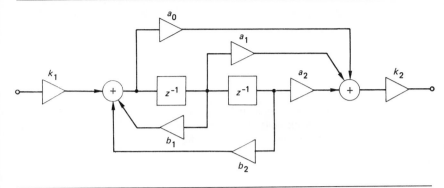

	Bandpass	Bandstop
$H_d(z)$	$\dfrac{k(z^2 - 1)}{z^2 - b_1 z - b_2}$	$\dfrac{k(z^2 + a_1 z + 1)}{z^2 - b_1 z - b_2}$
a_0	1	1
a_1	0	$-2b_1/(1 - b_2)$
a_2	-1	1
b_1	$2 \exp(-\omega_r T_s/2Q) \cdot \cos(\omega_d T_s)$	$2 \exp(-\omega_r T_s/2Q) \cdot \cos(\omega_d T_s)$
b_2	$-\exp(-\omega_r T_s/Q)$	$-\exp(-\omega_r T_s/Q)$
k_1	$(1 + b_2)/2$	$(1 + b_2)/2$
k_2	1	$(1 - b_2)/(1 + b_2)$
G_{max}	1.0	1.0

$\omega_r = 2\pi f_r = $ resonant frequency

$\omega_d = \omega_r \sqrt{[1 - 1/4Q^2]}$

Table 4.3 Second-order filter design parameters (impulse invariant)

Infinite attenuation will occur at the radian frequency ω_d. The width of the notch is determined by the Q of the original bandpass design. The transfer function for the notch filter is

$$H_d(z) = \frac{z^2 - 2\cos(\omega_d T_s) \cdot z + 1}{z^2 - 2\exp(-\alpha T_s) \cdot \cos(\omega_d T_s) \cdot z + \exp(-2\alpha T_s)}$$

The corresponding filter coefficients are given in Appendix 1 and the filter properties summarised in Table 4.4.

Comb filter Sometimes it is desirable to realize a filter with a series of notches equi-spaced along the frequency axis as would be required for example to remove mains harmonics. This type of filter is called a comb filter. We now know enough about discrete filters to design this directly. Zeros are obviously needed on the unit circle, spaced with an angular increment of $2\pi/k$ radians, where f_s/k is the frequency of the fundamental to be notched out. This will give infinite attenuation

TMS EXAMPLE No. 4.2

General second order filter section.

Begin main program.

```
START      IN      INPUT,ADC          :input signal

           LT      DEL1               :first delay element
           MPY     B1                 :feedback constant
           LTP     INPUT              :accumulate, set input scaling
           MPY     K1                 :scaling constant
           LTD     DEL2               :accumulate, load 2nd. delay
```

The 2nd. delay element is saved to a spare location DEL3 by the DMOV of the LTD instruction.

```
           MPY     B2                 :feedback constant
           LTD     DEL1               :accumulate, age first delay
           SACH    DEL1,1             :save new first delay
```

Note that the T register still holds the old first delay value.

```
           MPY     A1                 :feedforward constant
           LTP     DEL1               :accumulate, new delay to T
           MPY     A0                 :feedforward constant
           APAC                       :accumulate

           ADD     DEL3,15            :add 1.0*(2nd. delay)
```

Note that A2 is either +1.0 or −1.0, so the above line should be changed to SUB DEL3,15 when A2 is −1.0.

```
           SACH    RESULT,1           :save result
```

The final shift used to save the filter output is dependant on the value of the output scaling factor K2. Further multiplying and shifting may also be required at this point to implement K2 scaling.

```
           B       START              :back around
```

at the fundamental and subsequent harmonics. The width of the notches can now be controlled with poles at the same angle, but with magnitudes slightly less than unity.

The transfer function of the comb filter is thus

$$H_d(z) = \frac{z^k - 1}{z^k - \alpha^k}$$

where α is the magnitude of the poles.

Fig 4.18 gives an example. It is evident that although k delay sections are needed to realize the filter, there is only one feedforward and one feedback coefficient.

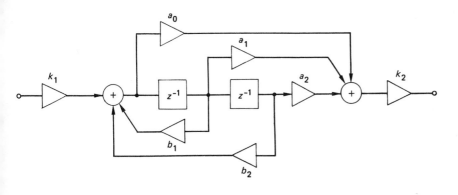

	Bandpass	Bandstop
$H_d(z)$	$\dfrac{k(z^2 - 1)}{z^2 - b_1 z - b_2}$	$\dfrac{k(z^2 + a_1 z + 1)}{z^2 - b_1 z - b_2}$
a_0	1	1
a_1	0	$-2b_1/(1 - b_2)$
a_2	-1	1
b_1	$-2(1 - \alpha^2 y)/(1 + \alpha x + \alpha^2 y)$	
b_2	$-(1 - \alpha x + \alpha^2 y)/(1 + \alpha x + \alpha^2 y)$	
k_1	$(1 + b_2)/2$	$(1 + b_2)/2$
k_2	1	$(1 - b_2)/(1 + b_2)$
G_{max}	1.0	1.0

$\omega_r = 2\pi f_r =$ resonant frequency
$\alpha = \omega_r \cot(\omega_r T_s/2) \approx 2/T_s$
$y = 1/\omega_r^2 \qquad x = 1/\omega_r Q$

Table 4.4 Second-order filter design parameters (bilinear transform)

A comb resonator, capable of selecting any periodic signal with fundamental f_s/k, can likewise be obtained if the zeros are all located at $z = 0$. This is illustrated in Fig 4.18c.

4.2.8 Low-pass to bandpass and high-pass transformations

Bandpass, bandstop and high-pass filter characteristics can all be synthesised from an equivalent low-pass transfer function using the appropriate transform procedure. These transforms are listed below for use with the discrete transfer function $H_d(z)$. Figs 4.19 and 4.20 illustrate the transformations.

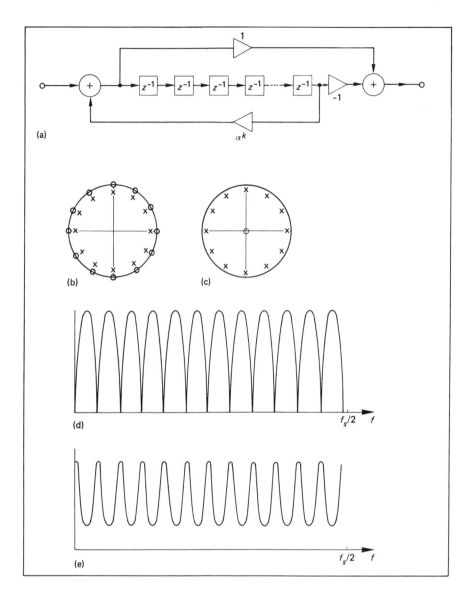

Fig 4.18 Comb filter
realization and response:
   a)   Filter realization
   b)   *z*-plane plot for notch
response
   c)   *z*-plane plot for
resonator response
   d)   Notch magnitude
response
   e)   Resonator magnitude
response

150

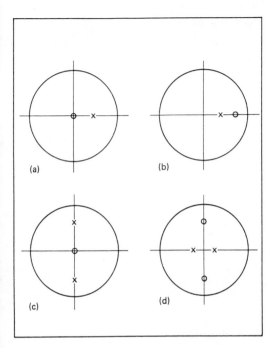

(a)

(b)

(c)

(d)

Fig 4.19   *z*-plane representation of filter transformations:
 a)   Low-pass prototype
 b)   High-pass transformation
 c)   Bandpass transformation
 d)   Bandstop transformation

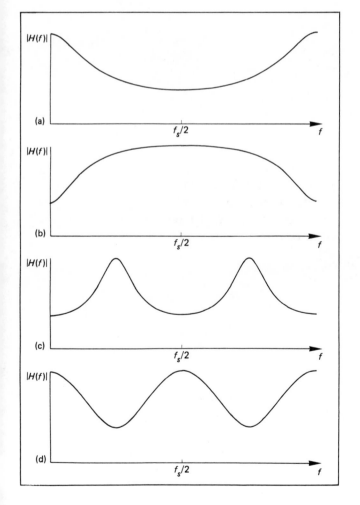

(a)

(b)

(c)

(d)

Fig 4.20   Magnitude response of filters defined in Fig 4.19:
 a)   Low-pass
 b)   High-pass
 c)   Bandpass
 d)   Bandstop

If the low-pass transfer function is first order, then the derived high-pass version is also first order with almost the same pole position, but the zero moved from $z = 0$ to very close to $z = 1$ (giving almost infinite attenuation at dc). The corresponding bandpass and bandstop transfer functions however are second order, with pole and zero positions close to those found using the direct synthesis technique outlined above.

High-pass transformation   (3 dB frequency $= f_c$)
Replace z in $H_d(z)$ by

$$z \to [\cos(2\pi f_c T_s) - z]/[1 - \cos(2\pi f_c T_s) \cdot z]$$

Bandpass transformation   (centre frequency f_0, 3 dB bandwidth $= f_c$)
Replace z in $H_d(z)$ with

$$z \to \frac{\alpha z - z^2}{-\alpha z + 1} \qquad \alpha = \frac{\cos(2\pi f_0 T_s)}{\cos(2\pi f_c T_s)}$$

Bandstop transformation   (centre frequency f_0, 3 dB bandwidth $= f_c$)
Replace z in $H_d(z)$ with

$$z \to \frac{z^2 - [2\alpha/(k+1)] \cdot z + (1-k)/(1+k)}{1 + [2\alpha/(k+1)] \cdot z + [(1-k)/(1+k)] \cdot z^2}$$

where $k = \tan^2(\pi f_c T_s)$ and $\alpha = \cos(2\pi f_0 T_s)/\cos(2\pi f_c T_s)$.

4.2.9   Summary and example

The techniques introduced in the previous sections for design of first- and second-order filter sections are the most common. Higher orders of filter are designed using exactly the same techniques, but usually with computer support. To summarise, the process of digital filter design and implementation can be divided into four main sections:

Evaluation of transfer function $H_d(z)$.
Selection of filter structure and subsequent
    coefficient evaluation.
Gain scaling for overflow protection.
Choice of algorithm implementation and
    coding of DSP program.

Finding $H_d(z)$
The main routes to finding $H_d(z)$ are either via existing analog filter design techniques and a suitable discrete transform, or by direct synthesis.

152

For the former, the bilinear transform is most widely used with suitable pre-warping of the analog design. The main alternative is the impulse response invariant technique, which results in a filter whose unit pulse response is a scaled sampled version of the desired continuous system impulse response.

The direct approach to digital filter design is either by judicious pole and zero placement in the z-plane, or using an FIR structure. FIR filters can be used to implement a sampled version of an arbitrary impulse response, including that of a linear phase filter. The penalty paid for the flexibility of FIR filters compared with IIR types is that greater memory storage and longer execution times are involved, although this disadvantage is partially offset by their inherent stability and insensitivity to coefficient quantization.

The difficulty of realizing any digital filter increases with the proximity of the poles to the unit circle. The pole location in the z-plane is dependent to a large extent on the relationship between the sampling frequency and the pole locations of the analog proto-type. In some cases it is possible to use a technique known as decimation to alleviate this problem, as discussed in section 4.7.

Choosing a filter structure
The structure needed to realize a given filter type will depend largely on the method used to derive $H_d(z)$. With the bilinear transform method, the structure shown in Fig 4.21 is appropriate. The filter is realized as a cascade of second-order blocks in preference to a single nth order system as it places less demands on coefficient accuracy.

Once the poles and zeros of $H_d(z)$ have been determined they should be mapped onto the second-order sections as follows:

Fig 4.21 Cascade realization of fourth-order filter from two second-order sections

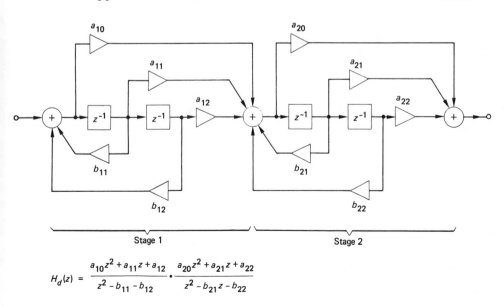

$$H_d(z) = \frac{a_{10}z^2 + a_{11}z + a_{12}}{z^2 - b_{11} - b_{12}} \cdot \frac{a_{20}z^2 + a_{21}z + a_{22}}{z^2 - b_{21}z - b_{22}}$$

153

a)   The highest Q pole pair and nearest zero pair are allocated to the last available section until all are allocated. If the prototype order is odd, an initial first-order section is needed. The relationship between filter coefficients and pole/zero values of the second-order section is given in Fig 4.21.

b)   In order to avoid internal overflow or excessive signal quantization, the input of each cascade section is scaled down or up respectively. This procedure is treated formally in section 4.6.

c)   A final gain correction multiplier is used where necessary.

There are various alternative methods for realizing a general second-order transfer function; however the method outlined above is the most widely adopted. There are also a number of different ways of combining second-order sections to realize higher-order filters. The parallel form (Fig 4.22) will be covered shortly, and the lattice-type structure is introduced in section 4.4.5 in connection with least squares design. In critical circumstances one configuration may give a better signal-to-noise ratio, or be less sensitive to coefficient quantization than another. The lattice structure has the advantage of inherent stability, despite the existence of coefficient quantization and is thus particularly convenient for adaptive filters.

When the design is based on impulse response invariance, it is convenient to use the parallel interconnection of first- and second-order blocks (Fig 4.22). The design method is described in detail in section 4.4.4.

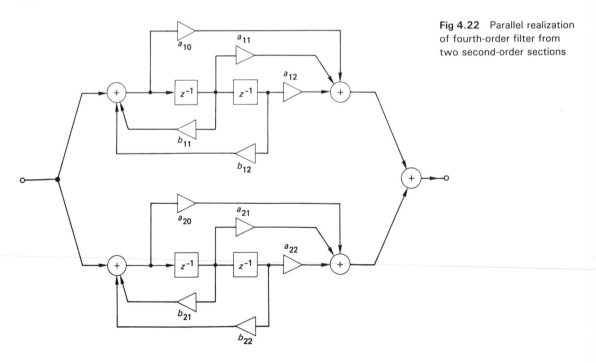

Fig 4.22   Parallel realization of fourth-order filter from two second-order sections

If $H_d(z)$ is realized by the direct method then either of the structures in Figs 4.21 and 4.22 may be used, assuming $H_d(z)$ has poles. In the all-zero case, the FIR structure of Fig 4.9 is appropriate.

Checking for overflow

In fixed-point implementations, with signal samples represented in Q15 format (cf. section 1.5), it is necessary to ensure that no output sample is ever greater than unity. This implies that the gain of the filter must itself not exceed unity (assuming a maximum input level of 1). In fact, an even more stringent condition is necessary. Consider the second-order IIR implementation in Fig 4.17. The signal at the input of the a_0 multiplier is the weighted sum of the input sample and the outputs of the two unit delay blocks. Depending on the feedforward coefficient values, it is clearly possible for overflow to occur at this point, even though the overall filter gain is less than unity. Thus, to be certain of avoiding overflow in a second-order IIR section it is necessary to evaluate the maximum gain at various summing points, and if this exceeds unity, to scale down the input to the filter accordingly. For a second-order section with z-plane poles at

$$z = \exp(-\alpha T_s) \text{ angle } \pm (\omega_d T_s)$$

the gain G_0 at the input summing point is approximately

$$G_0 = 1/[(1 - \exp(-\alpha T_s)) \cdot \exp(-\alpha T_s/2) \cdot 2 \, | \sin(\omega_d T_s) | \,]$$

Having ensured a maximum gain of unity to the summing point, the gain between the summing point and output must also be checked and, if necessary, the feedforward coefficients scaled to restore an overall gain of unity. Fortunately, this scaling task is ideally suited to CAD systems and is incorporated in the more recent filter design packages. (When there is a following stage in the cascade, the output scaling of the first stage can be combined with the input scaling of the second.) The way in which the scaling is distributed in a cascade has an important bearing on the overall 'signal to round-off' noise ratio, and also on the significance of coefficient quantization in the design. This issue is taken up in detail and illustrated with an example in section 4.6.

Code realization

Once the coefficient values have been derived, it is relatively straightforward to write the implementation code, particularly if the algorithm is first modelled and verified in a high-level language.

In Examples 4.3 and 4.4, a Pascal and TMS32010 version of an IIR filter are presented in which a cascade of second-order sections is realized by repeated calls to a subroutine which implements a general second-order filter section with up to two poles and two zeros. The same subroutine can realize a first-order filter by making the a_2 and b_2 coefficients zero. The delayed signal values are stored in an array (Pascal) or block of memory (TMS) and a separate three-element array or data memory area is used to pass the delay node signal values to the subroutine. The filter feedforward and feedback coefficients are stored in an array of successive memory locations and passed to the subroutine by means of an array index or pointer.

The actual filter being implemented in Example 4.4 is a 5th order elliptic design which is presented in detail in sections 4.4.3 and 4.6. The program has been structured to facilitate implementation of an IIR filter of arbitrary order, and as such does not represent the optimal solution in terms of execution time (which would avoid the use of subroutines and loops). Note in particular the manner in which coefficient values greater than unity are handled by either two-stage multiplication and accumulation and/or prescaling by a factor of two then left-shifting the result.

For completeness, an FIR filter implementation is illustrated in Example 4.5. In view of the symmetry of most FIR filter coefficients, a number of algorithms can be developed which minimise the number of multiplications at the expense of increased additions. When the execution time for a multiply and addition is identical, as for most current DSP devices, there is little benefit in this approach, and the algorithm used in Example 4.5 is recommended.

A summary of the various digital filter design strategies is given in Table 4.5.

Example filter design: an integrator
Suppose it is desired to use a DSP system to implement an integrator, i.e. $H_c(s)$ is required to approximate the function

$$H_c(s) = 1/s$$

as closely as possible over the range $f_s/10$ to $9f_s/10$.

In designing the transfer function $H_d(z)$, the following options are available:

1   The bilinear transform, which gives

$$H_d(z) = (T_s/2)(z+1)/(z-1)$$

2   The impulse response invariant method, which gives

$$H_d(z) = T_s \cdot z/(z-1)$$

1 *Determine transfer function $H_d(z)$ to meet filter specification*

IIR	FIR
Bilinear Transform*	Window Method*
Impulse Invariant*	Equiripple*
Least Squares	Frequency Sampling
Direct Method	Least Squares
	FFT

 *Available in commercial CAD packages

2 *Choose appropriate filter structure*

IIR	FIR
Parallel	Direct Form
Cascade	
Lattice	
Direct Form	

3 *Perform gain scaling where necessary*

IIR	FIR
Essential	Usually Unnecessary
Often Complex Process	Very Simple Process

4 *Define processing algorithm*

IIR	FIR
Difficult to Optimise	Very Straightforward

5 *Code generation*

IIR	FIR
Average Complexity	Very Simple

Table 4.5 Summary of digital filter design process

3   The direct method: the logical choice would be to place a pole at $z = 1$ and a zero at $z = -1$. (This is the same as the bilinear transform apart from a gain constant.)

All the above approaches would have to be rejected on two grounds. First they all involve realization of a pole on the unit circle which would lead to instability, and secondly none of the solutions gives the correct flat $-\pi/2$ phase shift over the range of frequencies of interest.

The only viable solution is to use the FIR method, which enables a band in the range $0 < f < f_s/2$ to be accurately approximated in both magnitude and phase. Details of the method are given in section 4.5.3.

EXAMPLE No. 4.3

Pascal implementation of an IIR filter.

Program IIR (input,output,indata1,outdata1);

IIR filter as cascade of second order sections. Input assumed to be in 'input.dat', 1024 samples long, output to 'output.dat'.

```
type
        arr20 = array [0..20] of real;
        arr2  = array [0..2]  of real;

var
        indata1,outdata1        : text;
        w,u,v                   : arr20;
        work                    : arr2;
        j                       : integer;
        x,y,acc,gain            : real;


Procedure secorder (pointer:integer; v,u:arr20; var w:arr20;x:real; var y:real);

var
        work                    : arr2;

begin
```

Accept input

```
        work[0] := x;
```

Load work area

```
        for j := 1 to 2 do
            work[j] := w[j+pointer];
```

Feedback term accumulation

```
        acc := 0;

        for j := 2 downto 0 do
            acc := work[j] * v[j+pointer] + acc;

        work[0] := 0;
```

Feedforward term accumulation

```
        y := 0;
        for j := 2 downto 0 do
            y := y + work[j] * u[j+pointer];
```

Shuffle data forward

```
        for j := 2 downto 0 do
            w[j+pointer+1] := work[j];

end; {secorder}
```

```
begin
```

Initialisation

```
        open (indata1,'input.dat',readonly);
        reset (indata1);
        open (outdata1,'output.dat');
        rewrite (outdat1);
```

Set up coefficient matrices

```
        gain := 0.5960509;

        u[0] := 1;
        u[1] := -0.51291382;        {first section feedforward}
        u[2] := 1.0;

        v[0] := 1;
        v[1] := 1.016917;           {first section feedback}
        v[2] := -0.81443393;

        u[3] := 1.0;
        u[4] := 0.93143255;         {second section feedforward}
        u[5] := 1.0;

        v[3] := 1;
        v[4] := 1.04061556;         {second section feedback}
        v[5] := -0.38125926;

        u[6] := 1;
        u[7] := 0;                  {third section feedforward}
        u[8] := 0;

        v[6] := 1;
        v[7] := 0.5;                {third section feedback}
        v[8] := 0;
```

Initialise state

```
        y := 0;

        for j := 0 to 5 do
            w[j] := 0;
```

Main loop

```
        for j := 0 to 1023 do
        begin
```

Input and output

```
            readln (indata1,x);
            writeln (outdata1,y,gain);
```

First section

```
            secorder (0,v,u,w,x,y);
            x := y;
```

Second section

```
                secorder (3,v,u,w,x,y);

            end;

        end. {IIR}
```

159

TMS EXAMPLE No. 4.4

Fifth order IIR filter implemented on TMS 32010 as cascade of second order sections. Passband edge at 0.1*(sampling frequency), stopband edge at 0.15*(sampling frequency). Passband ripple 0.3 dB, stopband attenuation 50 dB.

	AORG	0	:set address for program code
	B	INIT	:send resets to initialisation
	B	INTRPT	:send interrupts to handler

Begin main program.

Declare symbols used to reference constants and data memory locations.

WORK0	EQU	0	:delay node signals
WORK1	EQU	1	:for passing to subroutine
WORK2	EQU	2	
W01	EQU	3	:first stage delay node 0
W21	EQU	5	
W22	EQU	8	
W23	EQU	11	
.			
.			:space for higher orders
.			
POINTW	EQU	21	:pointers to arrays
POINTU	EQU	22	
POINTV	EQU	23	
OVERFL	EQU	24	:for coefficients > unity
INPUT	EQU	25	:offset binary input
XN	EQU	26	:2's complement input
OUTPUT	EQU	27	:offset binary output
YN	EQU	28	:2's complement result
ONE	EQU	29	:to hold 1
U01	EQU	30	:1st stage feedforward coefficients
U21	EQU	32	
U22	EQU	35	:second
U23	EQU	38	:third
V01	EQU	39	:1st stage feedback coefficients
V21	EQU	41	
V22	EQU	44	:second
V23	EQU	47	:third
C1	EQU	48	:first stage scaler coefficients
C2	EQU	49	:second
C3	EQU	50	:third

C4	EQU	51	:output scaler
UNITY	EQU	52	:unity coefficient
NCON	EQU	24	:number of values to initialise

Place lookup table in program memory for initialising data memory.

TABLE	DATA	1	:for ONE

U1 coefficients

	DATA	>4000	:0.5
	DATA	>4000	:0.5
	DATA	>0	:0

U2 coefficients

	DATA	>5312	:0.649
	DATA	>D9F5	:-0.2972
	DATA	>5312	:0.649

U3 coefficients

	DATA	>70D3	:1.882/2.1351
	DATA	>8001	:-2.1351/2.1351
	DATA	>70D3	:1.882/2.1351

V1 coefficients

	DATA	>7FFF	:1.0
	DATA	>5C84	:0.7228
	DATA	>0	:0

V2 coefficients

	DATA	>7FFF	:1.0
	DATA	>3A51	:1.4556 - 1
	DATA	>D464	:-0.6618

V3 coefficients

	DATA	>7FFF	:1.0
	DATA	>4113	:1.5084 - 1
	DATA	>F245	:-0.8928

Scaling coefficients

	DATA	>237B	:0.2772
	DATA	>1A65	:0.2062
	DATA	>1E42	:0.2364
	DATA	>08A5	:2.1351/2 - 1
	DATA	>7FFF	:1.0

Interrupt handler - interrupts not being used, so just a return.

INTRPT	RET		:return

Initialisation routine.

INIT	DINT		:disable interrupts
	SOVM		:overflow on
	LDPK	0	:use data page 0
	LACK	TABLE	:set source for data transfer
	LARK	AR0,ONE	:set destination
	LARK	AR1,NCON-1	:set number of values to transfer

```
RLOOP      LARP    AR0                    :point to AR0
           TBLR    *+,AR1                 :transfer value, update destination
           ADD     ONE                    :update source
           BANZ    RLOOP                  :repeat
```

Zero W array (delay nodes).

```
           LARK    AR0,W23                :point to first address
           LARK    AR1,8                  :do 9 times
           ZAC                            :clear accumulator

ZLOOP      LARP    AR0                    :point to AR0
           SACL    *-,AR1                 :store zero, update destination
           BANZ    ZLOOP                  :repeat
```

Now enter main loop.

```
START      IN      INPUT,PA0              :input signal

           LAC     INPUT
           SUB     ONE,15
           SACL    XN                     :save 2's complement input

           LACK    W21                    :set up for first stage
           SACL    POINTW
           LACK    U21
           SACL    POINTU
           LACK    V21
           SACL    POINTV
           ZAC                            :first stage V1 < 1
           SACL    OVERFL                 :hence overflow zero

           LT      C1
           MPY     XN                     :input passed in P register
           CALL    SECORD                 :perform first stage

           LACK    W22                    :set up for second stage
           SACL    POINTW
           LACK    U22
           SACL    POINTU
           LACK    V22
           SACL    POINTV
           LAC     UNITY                  :second stage V1 > 1
           SACL    OVERFL                 :hence overflow is 1

           LT      C2
           MPY     YN                     :previous stage output
           CALL    SECORD                 :perform second stage

           LACK    W23                    :set up third stage
           SACL    POINTW
           LACK    U23
           SACL    POINTU
```

```
        LACK    V23
        SACL    POINTV
        LAC     UNITY           :V1 again > 1
        SACL    OVERFL

        LT      C3
        MPY     YN
        CALL    SECORD          :perform third stage

        LT      C4              :output scaling
        MPY     YN
        PAC
        LT      UNITY           :allow for gain > unity
        MPY     YN
        APAC
        SACH    YN,1

        ZALH    YN              :use high accumulator
        ADDH    YN              :scale result up by 2
        SACH    YN              :save

        LAC     YN              :load result
        ADD     ONE,15          :adjust from 2's complement
        SACL    OUTPUT          :save offset binary value

        OUT     OUTPUT,PA0      :output result
```

A timing control routine should appear here to control the sampling frequency.

```
        B       START           :back for next input
```

Second order IIR section subroutine.

```
SECORD  LAR     AR0,POINTW      :load work area
        LARK    AR1,2

SUB11   LARP    AR0             :point to AR0
        LAC     *-,AR1          :load from source
        SACL    *               :store in work area
        BANZ    SUB11           :repeat

        PAC                     :new input in P register
        SACH    WORK0,1         :transfer to work area
```

Summation of feedback terms and input to form new delay node value.

```
        ZAC                     :clear accumulator
        LAR     AR0,POINTV      :set pointer
        LARK    AR1,2           :repeat count

SUB12   LARP    AR0
        LT      *-,AR1          :set T register
        MPY     *               :do multiply
        APAC                    :accumulate
        BANZ    SUB12           :repeat
```

```
          LT      OVERFL          :handles V1 > unity
          MPY     WORK1
          APAC
          SACH    WORK0,1         :update delay node 0
```

Summation of feedforward terms.

```
          ZAC                     :clear accumulator
          LAR     AR0,POINTU      :set pointer
          LARK    AR1,2           :source, and repeat count

          LARP    AR1

          LT      *-,AR0
          MPY     *-,AR1

SUB13     LTD     *-,AR0          :accumulate and shuffle
          MPY     *-,AR1          :multiply
          BANZ    SUB13           :repeat

          APAC
          SACH    YN,1            :new output

          LAR     AR0,POINTW      :restore from work area
          LARK    AR1,2           :source, and repeat count
          LARP    AR1

SUB14     LAC     *,AR0           :load from source
          SACL    *-,AR1          :store to destination
          BANZ    SUB14           :repeat

          RET                     :return from subroutine

          END                     :end of program
```

TMS EXAMPLE No. 4.5

Seventh order FIR filter, implemented on a TMS 32010.

```
        AORG    0                   :set address for program code

        B       INIT                :send resets to initialisation
        B       INTRPT              :send interrupts to handler
```

Begin main program.

Declare symbols used to reference constants and data memory locations.

```
W0      EQU     0                   :delay node signals
W7      EQU     7

INPUT   EQU     20                  :offset binary input
XN      EQU     21                  :2's complement input
OUTPUT  EQU     22                  :2's complement output
YN      EQU     23                  :offset binary output

ONE     EQU     30                  :to hold 1

U0      EQU     31                  :impulse response coefficients
U7      EQU     38

NCON    EQU     9                   :number of values to initialise
```

Place lookup table in program memory for initialising data memory.

```
TABLE   DATA    1                   :for ONE
```
U coefficients (Q15).
```
        DATA    >1000               :0.125
        DATA    >1000               :0.125
        DATA    >1000               :0.125
        DATA    >1000               :0.125
        DATA    >1000               :0.125
        DATA    >1000               :0.125
        DATA    >1000               :0.125
        DATA    >1000               :0.125
```

Interrupt handler - interrupts not being used, so just a return.

```
INTRPT  RET                         :return
```

Initialisation routine.

```
INIT    DINT                        :disable interrupts
        SOVM                        :overflow on
        LDPK    0                   :use data page 0

        LACK    TABLE               :set source for data transfer
        LARK    AR0,ONE             :set destination
        LARK    AR1,NCON-1          :set number of values to transfer
```

```
RLOOP     LARP    AR0         :point to AR0
          TBLR    *+,AR1      :transfer value, update destination
          ADD     ONE         :update source
          BANZ    RLOOP       :repeat
```

Zero W array (delay nodes).

```
          LARK    AR1,W7      :point to first address
          ZAC                 :clear accumulator
          LARP    AR1         :point to AR1

ZLOOP     SACL    *           :store zero
          BANZ    ZLOOP       :update destination, repeat
```

Now enter main loop.

```
START     IN      INPUT,PA0   :input signal

          LAC     INPUT       :offset binary format
          SUB     ONE,15
          SACL    XN          :save 2's complement input
```

Now perform FIR filtering.

```
          ZAC                 :clear accumulator
          LARK    AR1,W7      :point to final delay node
          LARK    AR0,U7      :point to corresponding coefficient
          LARP    AR1

          LT      *-,AR0      :set up first multiply
          MPY     *-,AR1
```

Now enter a loop to perform multiply, accumulate and data shuffle for the remaining delay nodes. Note that the auxiliary register AR1 is used both as a pointer to the delay nodes and as the loop counter, since the last delay node to be operated on (W0) is at address 0.

```
FLOOP     LTD     *,AR0       :add previous product, data shuffle
          MPY     *-,AR1      :perform current multiply
          BANZ    FLOOP       :adjust pointer and repeat

          APAC                :accumulate final product
          SACH    YN,1        :save result, remove extra sign bit

          LAC     YN          :load result
          ADD     ONE,15      :adjust from 2's complement
          SACL    OUTPUT      :save offset binary value

          OUT     OUTPUT,PA0  :output result
```

A timing control routine should appear here to control the sampling frequency.

```
          B       START       :back for next input
```

166

END   :end of program

COMMENTS.

In the above example, the filter is implemented using looped code, which is efficient in terms of program memory. Where execution time is a more important consideration, then in-line coding should be adopted.

The TMS 32020 has an enhanced instruction set, which includes single-instruction multiply, accumulate and multiply, accumulate, data shuffle operations. In combination with a repeat counter, these allow FIR filter taps to be performed in a single instruction cycle (see example 6.7).

4.3 Using filter design packages

For anything but the most basic first- and second-order filter realizations, the acquisition of a digital filter design package is to be recommended. Not only does it trivialise the task of filter coefficient derivation, but provides the flexibility and speed to select the appropriate filter type almost by trial and error. This is particularly useful when faced with limited DSP processing time for a given filtering task, and, perhaps more importantly, limited development time.

4.3.1 What packages can and cannot do

Filter design software packages will appeal immediately to all those engineers who have attempted to derive the transfer function $H_d(z)$ for a filter requirement by hand. Frequency selective filters, that is filters where the band from zero to $f_s/2$ Hz can be divided into passbands and stopbands, are far simpler to design with packages. All the user has to do is input the desired filter characteristics, and press the button! Gain characteristics are usually entered in the form of a mask, as shown in Figs 4.23 and 4.24 for the standard configurations of passbands and stopbands. In the passbands the nominal transfer function magnitude is taken to be unity, but in order to ease the filter order required, variations in magnitude from $1 + \delta_p$ to $1 - \delta_p$ are permitted. This corresponds to an allowable passband ripple R_{pb} of

$$R_{pb} = 20 \log_{10}[(1 + \delta_p)/(1 - \delta_p)] \text{ dB}$$

In the stopband, the transfer magnitude is nominally zero but magnitudes up to δ_s are tolerated, corresponding to a minimum stopband attenuation R_{sb} of

$$R_{sb} = 20 \log_{10}(\delta_s) \text{ dB}$$

167

Since sharp transitions between passbands and stopbands are also impossible with finite-order filters, it is necessary to have a finite *transition region* between each pass and stopband. This is normally accommodated by specifying separate frequencies for the edge of the passband and the edge of the stopband, f_{pb} and f_{sb} respectively.

The package derives an $H_d(z)$ which complies with the mask, using one of the methods outlined in the previous section. In the case of bandpass, bandstop and high-pass filters the mask would first be transformed into the equivalent low-pass mask, the low-pass $H_d(z)$ determined, and finally $H_d(z)$ transformed into the appropriate form using the techniques explained in section 4.2.8—all in a manner entirely transparent to the user. Most packages also contain routines for housekeeping tasks such as the evaluation and display of

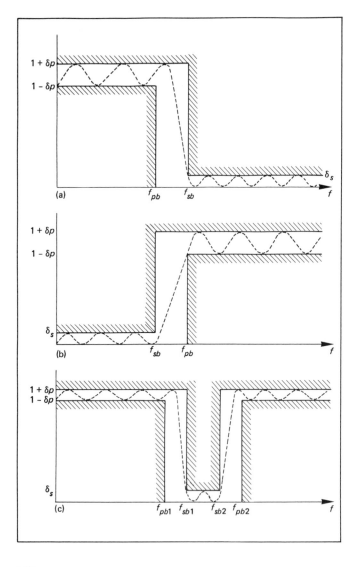

Fig 4.23 Specification of filters by means of masks:
  a) Low-pass
  b) High-pass
  c) Bandstop

magnitude and phase characteristics from $H_d(z)$, calculating the z-plane poles and zeros from the feedforward and feedback coefficients and vice-versa, and for converting (via the Fourier transform) from the transfer function to the impulse response.

Usually the designer wishes to develop the simplest filter that is capable of performing the task at hand. The packages allow 'what if' experimentation to assess the sensitivity of filter order to various modifications of the specification mask. In section 4.3.3 some general rules covering the nature of these trade-offs are given.

Although good for designing frequency selective filters and certain other well defined responses such as differentiators and Hilbert transform filters, filter software packages only rarely support the design of filters, such as equalizers, where an arbitrary magnitude

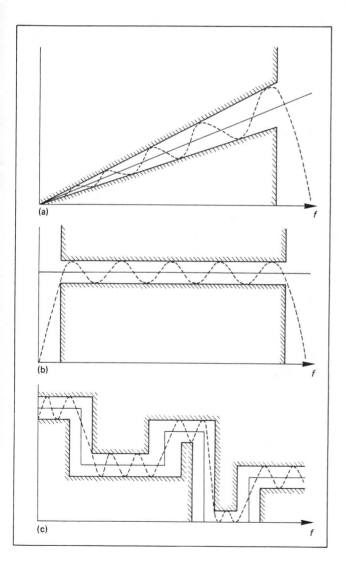

Fig 4.24 Standard masks for:
    a) Differentiator
    b) Hilbert transform filter
    c) Multiple band filter

and phase characteristic is to be approximated. Fortunately it is not difficult to write software to support this type of filter design, as is demonstrated in section 4.5. The key routine required is the fast Fourier transform, discussion of which is postponed until Chapter 5.

Apart from the filter specification, the only other input parameter to the design program is the choice of design method. The first choice is between IIR and FIR designs; then within each design type there is usually a choice of design methods. The next section gives some guidance as to which to choose, with more detail being given in sections 4.4 and 4.5.

If fixed-point processors are to be used for implementation, the next task is to evaluate the filter response with finite accuracy coefficients. If the response no longer meets the mask, a second design iteration may be needed using a tighter mask. Some packages allow the word length of the coefficients to be specified during the design phase.

It should be emphasised that the design technique upon which most packages are based is that of finding the filter order required to satisfy a specified mask. An alternative approach is to specify the order of the filter and find the coefficients which give the best fit, in, say, a least squares sense, to the desired response. Design on this basis is discussed in section 4.4.5.

4.3.2 Selecting the filter type and design method

The following provides some simple rules for choosing the filter type and design method. (Note that simple first- and second-order filters can be designed directly from the coefficient values given in Appendix 1.)

1 If perfect *linear phase* is needed, the filter must be FIR. (Phase equalization of IIR filters is possible for less precise phase requirements.) For all other specifications either FIR or IIR designs can be used. An IIR filter will generally be much less demanding on processor resources but is more difficult to implement and may be subject to instability when the transition width is a small fraction of $f_s/2$ (due to coefficient quantization sensitivity).

2 For a *frequency selective magnitude response* with an unspecified phase response, an IIR filter using the bilinear transform of an elliptic analog prototype will give the most efficient design which meets the specification. The Chebyshev approximation should be used if either the pass or stop-band must be monotonic (ripple free), and the Butterworth approximation used if both pass and stop-bands must be monotonic. Since analog prototypes only exist for four basic frequency selective characteristics (low-pass, high-pass, bandpass and bandstop) only these can be designed in this way.

3 For *linear phase response frequency selective filters*, including those with multiple pass and stop-bands with different gains and permitted ripple, an FIR filter designed with the equiripple (also called Parks-McClellan) technique, based on the Remez exchange algorithm, should be used. The permitted order of the filter may be limited by the computation time required for its design, in which case one of the other FIR methods must be adopted.

4 For *linear phase frequency selective filters* where the stopband attenuation is required to increase as the distance from the passband edge increases, an FIR filter based on the window method gives the user greatest flexibility in trading off the rate of roll-off of the response against ultimate stopband attenuation. It is not possible to independently control pass and stopband ripple with this method, and the order required to realize a narrow transition will usually be larger than that yielded by the equiripple method.

5 Either the equiripple or the window methods of FIR filter design can be used for differentiators or Hilbert transform filters.

6 To approximate arbitrary impulse responses or a set of magnitude and phase characteristics, FIR filters designed using the frequency sampling method are called for. The only support needed for this method is a fast Fourier transform routine and a routine for evaluating the resulting filter frequency response of $H_d(z)$ from its zeros (see section 2.7). The method is sub-optimal in the sense that a lower-order filter could be designed to achieve a given mask using one of the other FIR methods, and it should therefore only be used when the desired response cannot be specified as a *fixed set* of pass and stopbands.

These factors are summarised in Table 4.6.

Filter properties	Filter type	
	IIR	FIR
Linear Phase	NO (improved with equalization)	YES
Hilbert Transform	YES (approximation only	YES
Differentiator	YES (approximation only)	YES
Filter Order for Given Gain Response	LOW	HIGH
Memory Requirement	LOW[*]	HIGH[*]
Execution Time	LOW[*]	HIGH[*]
Algorithm Complexity	HIGH	LOW
Stability	GOOD (with careful design)	EXCELLENT
Sensitivity to Word Length	HIGH (satisfactory with 16-bit coefficients)	LOW
	[*]Assumes identical gain response specification.	

Table 4.6 Summary of FIR and IIR filter characteristics

4.3.3   Choice of filter specification

The order of the digital filter needed to approximate a given response can depend in quite subtle ways on the nature of the specification. The most important general rule is that the filter order will be set by the sharpest transition, measured in terms of the ratio:

$$\frac{\text{Width of the transition region in Hz}}{f_s/2 \text{ in Hz}}$$

It is important to realize the influence of the sampling frequency on the order of digital filter required. A modest analog filter characteristic can well involve a very high order digital filter realization if the cut-off frequency is small compared with the sampling rate. If this is the case, a reduction in the sampling rate should be considered. Section 4.7 describes how different filters can be implemented with different sampling rates yet make use of the same DSP device, by using the process of discrete sub-sampling and re-sampling known as decimation and interpolation.

High-pass and bandstop FIR filters, where the passband includes $f_s/2$, can only be realized using an *odd number of taps*, whereas Hilbert transform filters and differentiators which have a high-pass frequency response must have an *even number of taps*.

The number of non-zero coefficients needed to realize a given-order FIR filter can be cut in half by the simple expedient of arranging the mask to have even symmetry about $f_s/4$. For bandpass and bandstop filters this means that the filter must be centred on $f_s/4$ and have *equal weighting* of the passband and stopband ripple. For low-pass and high-pass filters, the transition region must be centred on $f_s/4$. This results in every second sample of the impulse response being zero. The technique is only applicable for filters with *odd* numbers of taps.

These filter characteristics are summarised in Table 4.7.

4.3.4   Two example packages

DFDP (Atlanta Signal Processors Incorporated)
This is an IBM Personal Computer based package with four program modules:

IIR	A program for designing recursive filters by the bilinear transform method, using Butterworth, Chebyshev or Elliptic prototypes (up to 20th order).
KFIR	A program for designing FIR filters by the method of windowing, using a Kaiser window (up to 510 taps).

172

Filter type	Filter property for linear phase
Low-pass	Odd or even filter lengths allowed. Impulse response exhibits even symmetry.
High-pass	Odd filter lengths only if passband includes $f_s/2$. Impulse response exhibits even symmetry.
Bandpass	Odd and even filter lengths allowed. Impulse response exhibits even symmetry.
Bandstop	Odd filter lengths only if passband includes $f_s/2$. Impulse response exhibits even symmetry.
Hilbert Transform	Odd and even filter lengths allowed; even lengths only if passband includes $f_s/2$. Gain must fall to zero at dc. Impulse response exhibits odd symmetry.
Differentiator	Odd and even filter lengths allowed; even lengths only if passband includes $f_s/2$. Gain must fall to zero at dc. Impulse response exhibits odd symmetry.

For all linear phase FIR filters, making the gain response symmetrical about $f_s/4$ forces every second coefficient to zero (passband and stopband ripple must be identical). Such filters are called 'half-band' filters.

Group Delay $\tau = (N-1) \cdot T_s/2$

Table 4.7   Summary of FIR filter properties

PMFIR   A program for designing optimal FIR filters by the Remez exchange algorithm developed by Parks and McClellan (up to 130 taps).

CGEN   A program for generating assembly language code for the TMS320 family of DSP devices to implement filters designed by the above modules.

Assuming that one of the standard frequency selective filters is required, the chosen program module prompts the user for the specification mask in terms of passband and stopband edge frequencies, the passband and stopband tolerance parameters δ_p and δ_s and the sampling frequency. The program then calculates an estimate of the filter order required (it will be exact in the IIR case if the realization is floating point). At the user's request the design is then carried out and the actual frequency response (taking into account finite word length) is calculated and compared with the mask. If the specifications are not met, the user is given the option of increasing the filter order and attempting a redesign. On completion the user may choose to display one or more of the following:

Magnitude response.
Log magnitude response.
Phase response.
Group delay (derivative of phase response).

z-plane pole/zero plot.

Impulse response.

and obtain a hard copy of both the graphics display and the design itself. The IIR filters are implemented as a cascade of second-order sections and the FIR filters as a standard transversal structure.

If the program selected is an FIR filter design procedure, then multiband filters, differentiators and Hilbert transform filters can be designed in addition to the standard frequency selective types. Fig 4.24 illustrates the nature of the masks involved. Note how the passband tolerance in the case of the differentiator is a fixed proportion of the passband gain, and how no transition band near $f_s/2$ is needed when the order of the filter is odd, as explained earlier.

The Design Example on page 175 is an IIR design produced by DFDP using program module IIR and an elliptic approximation. Both the design mask and the final response are plotted. Note that the feedforward and feedback coefficients for the ith cascaded second-order section are labelled as $B(i,k)$ and $A(i,k)$ respectively where k is the associated number of delays. These correspond to a_{ik} and $-b_{ik}$ in the notation of Fig 4.21.

The package may be purchased directly from ASPI, 770 Spring St, N.W., Suite 208, Atlanta, Georgia 30308, USA or through local agents.

IEEE Programs for Digital Signal Processing
This collection of published programs, written in ANSI standard Fortran IV, is available on magnetic tape for a nominal cost from IEEE, 445 Hoes Lane, Piscataway, N.J. 08854 USA. The collection is very comprehensive, covering not only the standard digital filter design methods but a number of general-purpose programs for the fast Fourier transform, correlation, convolution and spectral analysis (these topics are covered in Chapter 5 of this book). There is no standard interface for transfer of data between programs and some work is required to adapt each program to its host installation. Table 4.8 summarises the programs which comprise the IEEE set. Many of the programs are also available in the open literature (Ref 4.5).

4.4 IIR filter design

Infinite Impulse Response filters are usually the best choice when a non-linear phase characteristic can be tolerated, i.e. when only the magnitude characteristic is of interest. Possible exceptions to this general rule are designs where sensitivity to the quantization of the coefficients is a problem and cases where there is sufficient processing

DESIGN EXAMPLE

The design module IIR was used to determine that a particular set of specifications for a bandpass filter could be met with a l6th-order Butterworth, a l0th-order Chebyshev, or an 8th-order elliptic recursive filter. The following gives the specifications and the approximation errors for the elliptic filter:

*** CHARACTERISTICS OF DESIGNED FILTER ***

ELLIPTIC BANDPASS FILTER

FILTER ORDER = 8
SAMPLING FREQUENCY = 10.000 KILOHERTZ

	BAND 1	BAND 2	BAND 3
LOWER BAND EDGE	.00000	2.00000	3.50000
UPPER BAND EDGE	1.50000	3.00000	5.00000
NOMINAL GAIN	.00000	1.00000	.00000
NOMINAL RIPPLE	.01000	.01000	.01000
MAXIMUM RIPPLE	.00518	.00568	.00518
RIPPLE IN DB	-45.71958	.04918	-45.71628

The filter coefficients were quantized to 16 bits for a cascade-form implementation of the filter. In the following table of coefficients printed by the program, the $B(I,J)$ coefficients determine the zeros, and the $A(I,J)$ coefficients determine the poles of the filter.

I	A(I,1)	A(I,2)	B(I,0)	B(I,1)	B(I,2)
1	.281708	.652527	.234985	.411163	.234650
2	-.281677	.652527	.278381	-.487091	.277969
3	.643402	.877625	.423798	.521332	.423676
4	-.643372	.877625	.664490	-.817413	.664307

The following lineprinter hardcopy shows the log magnitude in dB for the band from 0 to 5 kHz and an exploded view of the passband of the filter.

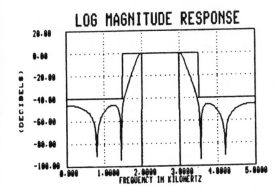
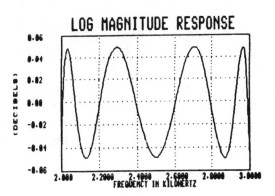

Numerical Transform Programs
    General-purpose FFT
    Special-purpose FFTs
    Mixed radix FFT
    Chirp-z transform
    Winograd Fourier transform
    Two-dimensional FFT

Power Spectrum Analysis and Correlation
    Power spectrum via periodogram
    Power spectrum via correlation
    Coherence and cross-spectrum

Fast Convolution (FFT filtering)

Linear Prediction Speech Analysis

Filter Design
    Parks-McClellan FIR
    Window method FIR
    Finite word length FIR subroutine
    IIR design (4 programs)

Spectrum Analysis

Interpolation and Decimation
    Interpolator design
    Sampling rate conversion
    Multistage decimation/interpolation/
        narrow band filtering

Table 4.8   Programs available in the IEEE set

capacity and it is desired to exploit the unconditional stability and ease of design of FIR filters. All IIR filters have z-plane poles as well as zeros and are realized with a recursive structure. The most common design techniques are: (i) the bilinear transform method which derives a suitable $H_d(z)$ from an analog transfer function with known properties; (ii) the impulse response invariant method; (iii) direct pole and zero placement; and (iv) derivation of the pole locations by least squares fit to a desired characteristic. Methods (i), (ii) and (iv) are considered in this section, (iii) having been adequately covered in section 4.2.7.

The bilinear transform method is very convenient when the filter specification is one of the four standard types of frequency selective response. The impulse response invariant method is used when the equivalent analog system impulse response $h_c(t)$ is to be identical with a specified minimum phase impulse response at the sampling moments. (In this case the frequency response will not match the specified system due to aliasing.) Direct pole and zero placement is convenient for the realization of simple first-order responses and resonators. Finally, the least mean square approach is used when the best approximation is required for a specified order of filter.

4.4.1 The bilinear transform

This is a method for designing a discrete transfer function $H_d(z)$ that will have the same general characteristics as an analog prototype transfer function $H_p(s')$. To be more precise it will have the same arrangement of pass and stopbands, the same passband ripple and stopband attenuation, the only difference being the frequency at which a feature in the response occurs. This is because the continuous system frequency response spans the range zero to infinity, whereas in the discrete filter the same magnitude variations must fit into the span zero to $f_s/2$ (Fig 4.6). The term *frequency warping* is used to describe this process.

Note the use of s' for the complex frequency variable in the warped prototype filter to distinguish it from s which is the complex frequency variable for the actual continuous transfer function $H_c(s)$ to be realized by the DSP system, incorporating the discrete filter $H_d(z)$.

The advantage of the bilinear transform method is that it enables the wealth of existing analog filter design theory to be applied directly to the design of discrete filters. For a given filter type and order it is possible to calculate the exact passband ripple, transition band width and minimum stopband attenuation. Similarly, for a specified mask and filter type it is possible to accurately assess the order of filter required to meet the given specification. The steps involved in the design of $H_d(z)$ using the bilinear transform are as follows:

a)   Specify $H_c(s)$ in the form of a mask.
b)   Derive the mask for $H_d(z)$.
c)   Find the equivalent mask for $H_p(s')$ (predistortion).
d)   Find the poles and zeros of $H_p(s')$ using one of the standard approximations (Butterworth, Chebyshev, Elliptic).
e)   Derive $H_d(z)$ from $H_p(s')$.
f)   Adjust the dc gain if necessary.

An example filter design is given in section 4.4.3.

We now derive the relationship between $H_d(z)$ and the associated analog prototype $H_p(s')$. For exact equivalence between the actual filter realization $H_c(s)$ and the analog prototype $H_p(s')$ we require

$$H_c(s) = H_d(z)\big|_{z=\exp(s \cdot T_s)} = H_p(s')$$

or

$$H_p(s')|_{s' = (1/T_s)\log_e z} = H_d(z)$$

Unfortunately this does not give rise to a rational, thus realizable $H_d(z)$. If however $\log_e z$ is approximated by the expansion:

$$\log_e z = 2(z - 1)/(z + 1)$$

then the $H_d(z)$ derived by replacing s' with $(2/T_s)(z - 1)/(z + 1)$ is indeed realizable. The transformation is in fact a one-to-one mapping of the entire imaginary axis in the s'-plane onto the unit circle in the z-plane (Fig 4.25). Consequently there is a point on the unit circle where the magnitude and phase response of $H_d(z)$ will be identical with that of the analog prototype for an arbitrarily chosen frequency f'. The exact mapping is

$$j2\pi f' = (2/T_s)(z - 1)/(z + 1)$$

hence

$$z = (1 + j\pi f' T_s)/(1 - j\pi f' T_s)$$

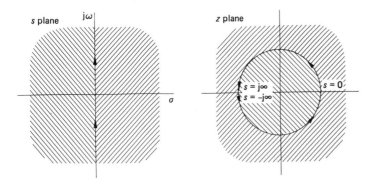

Fig 4.25 Bilinear transform mapping of s-plane into z-plane

To find the frequency f in the digital filter response which corresponds to f' in the analog prototype, z can be expanded to give

$$z = \cos(2\pi f T_s) + j \sin(2\pi f T_s) = (1 + j\pi f' T_s)/(1 - j\pi f' T_s)$$

Solving this, using the trigonometric identity

$$\tan 2\phi = 2\tan\phi/(1 - \tan^2 \phi)$$

gives

$$f = (1/\pi T_s)\arctan(\pi f' T_s)$$

To use the bilinear transform method the band edges of the required $H_d(z)$ are mapped to the equivalent analog prototype frequencies f' using

$$f' = (1/\pi T_s)\tan(\pi f T_s)$$

From the known properties of one of the standard filter approximations (Butterworth, Chebyshev, etc), the poles and zeros of $H_p(s')$ needed to satisfy the f' mask are now derived, either from a book of filter tables or from a filter design software package. Finally, using the substitution:

$$s' = (2/T_s)(z - 1)/(z + 1)$$

the discrete transfer function is yielded. If $s' = \sigma$ is a pole or zero in $H_p(s')$, then $H_d(z)$ has a corresponding pole or zero at

$$z = [1 + (T_s/2)\sigma]/[1 - (T_s/2)\sigma]$$

and there will also be a zero or pole at $z = -1$, depending upon whether σ was a pole or zero respectively. If necessary the low-pass $H_d(z)$ can be transformed into one of the other standard types (bandpass, high-pass, bandstop), as outlined in section 4.2.8. A further example can be found in section 4.4.3.

4.4.2 Choice of filter approximation

Brief notes on the three major filter approximations are given below and are illustrated by means of a practical design specification presented in Figs 4.26 to 4.29. Each filter satisfies the same low-pass tolerance mask, which has approximately ± 0.42 dB of passband ripple, 26 dB minimum stopband attenuation ($\delta_p = \delta_s = 0.05$), and a transition band from $0.15 f_s$ to $0.2 f_s$.

To comply with the mask, the required filter orders are

Butterworth	11
Chebyshev	6
Elliptic	4

A fourth approximation (Bessel) is commonly used in analog filter design when there is a requirement for the phase characteristic to be as linear as possible. As we have seen, perfectly linear phase characteristics can be achieved with FIR digital filters in any case, so for digital filters the Bessel approximation has little relevance.

The Butterworth approximation
An nth order Butterworth low-pass approximation has the property that the first n derivatives of the transfer function magnitude are zero at zero frequency. The magnitude response decreases monotonically from zero with ever-increasing slope, asymptotically approaching $-6n$ dB per octave. The poles of a 1 radian/second low-pass Butterworth transfer function are uniformly spaced around the unit circle in the left-half of the s-plane, at angular intervals of π/n radians. There are no zeros. Of all the approximations we shall

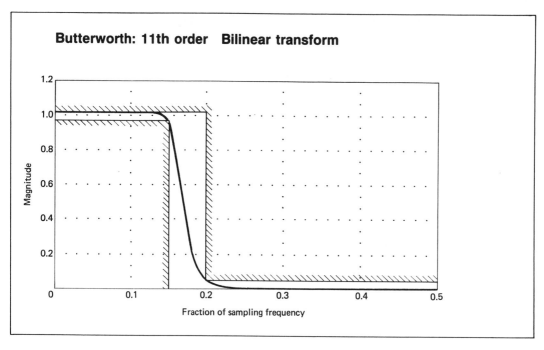

Fig 4.26 Butterworth filter design for specified mask

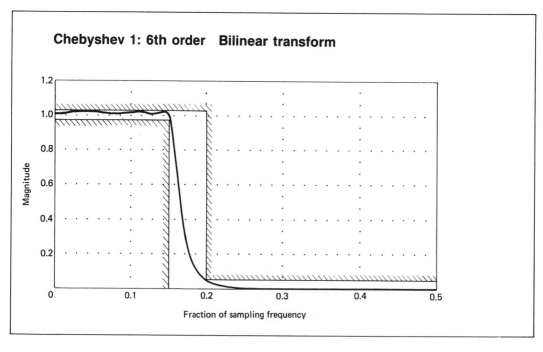

Fig 4.27 Chebyshev type I design for specified mask

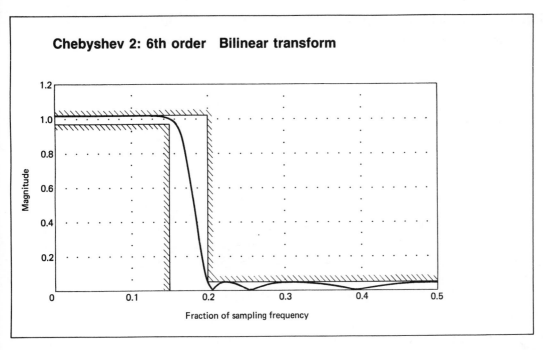

Fig 4.28 Chebyshev type II design for specified mask

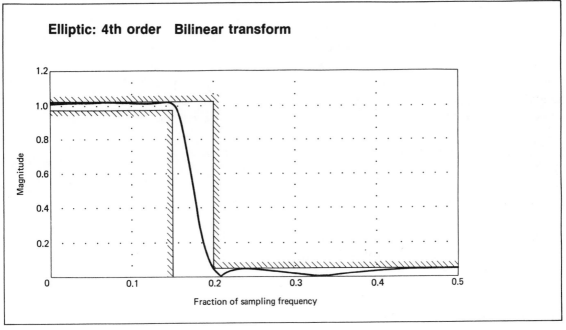

Fig 4.29 Elliptic design for specified mask

181

consider, Butterworth filters require the highest order to comply with a given tolerance mask, and they should only be used when the inherent smoothness of the response is essential. Note that, because of frequency warping, the discrete version of the Butterworth characteristic has a steeper roll-off for the same order than the analog version, and that as well as the n z-plane poles there are n zeros at $z = -1$.

The Chebyshev approximation
In the Chebyshev approximation the characteristic is monotonic (ripple free), either in the passband or the stopband, but not in both. *Type I* filters have an equiripple passband and *type II* an equiripple stopband. Type I is an all-pole continuous transfer function in which the imaginary parts of the poles are the same as in the Butterworth design for the same order, but the real parts of the poles are closer to the imaginary axis. For a passband tolerance δ_p and a stopband tolerance δ_s, the filter order required can be derived from the expression

$$n = \frac{\cosh^{-1}(1/\epsilon\delta_s)}{\cosh^{-1}(f_{sb}/f_{pb})}$$

where $\epsilon = 1/(1 - \delta_p)^2 - 1$.

Both type I and type II approximations have $n/2$ pole pairs. The type II also has $n/2$ zero pairs located in the stopband region on the imaginary axis.

The Elliptic (or Cauer) approximation
Elliptic designs are most efficient in the sense that they enable the realization of a given mask with the minimum filter order. Both pass bands and stopbands have an equiripple characteristic. The analog prototype has $n/2$ s-plane pole pairs and $n/2$ imaginary axis zero pairs.

The pertinent filter properties are summarised in Table 4.9.

4.4.3 Design example: fifth-order elliptic filter

Suppose it is desired to realize an elliptic, discrete time, low-pass filter which complies with the following tolerance mask:

passband edge	1000 Hz
stopband edge	1500 Hz
sampling rate	10 000 Hz
passband ripple	± 0.3 dB ($\delta_p \approx 0.035$)
stopband attenuation	50 dB ($\delta_s \approx 0.00316$)

Filter type	Filter property
Bessel	Smooth gain response.
	Good (almost linear) phase response.
	Very poor attenuation vs. filter order.
	(FIR filters often used in preference)
Butterworth	Smooth gain response.
	Reasonably good phase response.
	Poor attenuation vs. filter order.
Chebyshev	Type I: Ripple in passband—smooth stopband.
	Type II: Ripple in stopband—smooth passband.
	Poor phase response.
	Good attenuation vs. filter order.
Elliptic (Cauer)	Ripple in passband and stopband.
	Very poor phase response.
	Excellent attenuation vs. filter order.
	Optimum choice when phase response not critical.

Table 4.9 Summary of analog filter properties

By predistorting the band edge frequencies using

$$f' = (1/\pi T_s)\tan(\pi f T_s)$$

the mask of the analog prototype is derived:

passband edge	1034.2515 Hz
stopband edge	1621.8698 Hz
sampling rate	10 000 Hz
passband ripple	± 0.3 dB
stopband attenuation	50 dB
f_{sb}/f_{pb}	1.568

Using published data (Ref. 4.1), an elliptic filter with order 5 has a passband ripple of 0.28 dB, a minimum stopband attenuation of 50.1 dB and a transition band ratio of 1.556, thus satisfying the mask. The poles and zeros of a 5th order low-pass elliptic transfer function as given in the tables are, for a cut-off frequency of 1 radian/sec:

pole pair 1	$s' = -0.09699 \pm j1.0300$
pole pair 2	$s' = -0.33390 \pm j0.7177$
real pole	$s' = -0.49519$
zero pair 1	$s' = \pm j1.6170$
zero pair 2	$s' = \pm j2.4377$

Scaling the above poles and zeros by the factor $2\pi f_{pb} = 6498.39$ to achieve the desired filter cut-off frequency results in the poles and zeros of $H_p(s')$. These are now converted to the poles and zeros of $H_d(z)$ by replacing s' with $(2/T_s)\,(z-1)/(z+1)$:

pole pair 1 $z = 0.9449$ angle $37.04°$
pole pair 2 $z = 0.8135$ angle $26.5°$
real pole $z = 0.7228$

zero pair 1 $z = 1.0$ angle $55.44°$
zero pair 2 $z = 1.0$ angle $76.76°$
real zero $z = -1.0$

The real zero at -1.0 results from the unit imbalance between the number of poles and zeros in $H_p(s')$ since each pole in $H_p(s')$ generates a zero at $z = -1$ in $H_d(z)$ and vice versa.

The filter is realized with a cascade consisting of a first-order section followed by two second-order sections in accordance with the noise-limiting strategy outlined in section 4.2.7. The choice of scaling factors C_1, C_2 and C_3 is left to section 4.7 where the example is taken up again and followed through to yield the actual feedforward and feedback coefficients for each of the stages in the cascade.

The frequency response for the example filter is given in Fig 4.30.

4.4.4   Design by impulse response invariance

To realize a design which matches the impulse response of an nth order continuous transfer function $H(s)$, it is necessary to expand $H(s)$ in partial fractions as shown below. This is only possible if the number of poles is at least one more than the number of zeros. The method is illustrated with a transfer function which has i complex conjugate zero pairs and j complex conjugate pole pairs, where $i < j$. If

$$H(s) = \frac{k(s-z_1)(s-z_1{}^\star)\ldots(s-z_i)(s-z_i{}^\star)}{(s-p_1)(s-p_1{}^\star)\ldots(s-p_j)(s-p_j{}^\star)}$$

then the partial fraction expansion is

$$H(s) = \frac{k_1}{(s-p_1)} + \frac{k_1{}^\star}{(s-p_1{}^\star)} + \cdots + \frac{k_j}{(s-p_j)} + \frac{k_j{}^\star}{(s-p_j{}^\star)}$$

where $k_n = H(s)\cdot(s-p_n)\big|_{s=p_n}$

Each conjugate pole pair in the partial fraction expansion can be represented in the time domain as an exponentially decaying sinusoidal component in the impulse response. If $p_n = \alpha + j\omega_d$ then the impulse response associated with the nth pole pair is

$$h_n(t) = 2\,\mathrm{Re}(k_n)\cdot\exp(-\alpha t)\cdot\cos(\omega_d t)$$
$$\qquad - 2\,\mathrm{Im}(k_n)\cdot\exp(-\alpha t)\cdot\sin(\omega_d t)$$

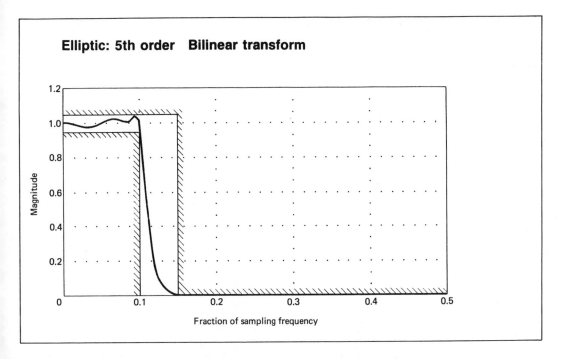

Elliptic: 5th order   Bilinear transform

Each such component is sampled, *z*-transformed and realized by one of the second-order parallel blocks in the structure of Fig 4.22, with the coefficients given in Table 4.10. It is of course possible to find the equivalent $H_d(z)$ for the serial connection of second-order discrete processes and thus use the structure in Fig 4.21. Equally well, a discrete transfer function designed by the bilinear transform method can be resolved into partial fractions and realized as in Fig 4.22. The issues in the choice between the two structures are discussed in section 4.6.

Fig 4.30 Elliptic filter design example

When there is one or more real poles, then the partial fraction expansion will include an equivalent number of terms of the form

$$k_i \exp(-\alpha_i t)$$

where k_i is real. The parallel realization will then include an

---

Analog second-order impulse response
$$h(t) = 2\mathrm{Re}(k_i) \cdot \exp(-\alpha t) \cdot \cos(2\pi f_d t) - 2\mathrm{Im}(k_i) \cdot \exp(-\alpha t) \cdot \sin(2\pi f_d t)$$

Coefficients in discrete realization of Figure 4.17
$$a_0 = 2 T_s \mathrm{Re}(k_i)$$
$$a_1 = -2 T_s \exp(-\alpha T_s)[\mathrm{Re}(k_i) \cdot \cos(2\pi f_d T_s) - \mathrm{Im}(k_i) \cdot \sin(2\pi f_d T_s)]$$
$$b_1 = 2 \exp(-\alpha T_s) \cdot \cos(2\pi f_d T_s)$$
$$b_2 = -\exp(-2\alpha T_s)$$

Table 4.10 Discrete system coefficients to realize an analog second-order impulse response at the sampling instants

Table 4.11 Discrete system coefficients to realize an analog first-order impulse response at the sampling instants

equivalent number of first-order sections designed as indicated in Table 4.11.

4.4.5 Least squares design

If a discrete transfer function $H_d(z)$ is all-pole, then a complementary all-zero (FIR) filter can be designed such that when the two are cascaded the combined system has a unit pulse response. For example, if

$$H_d(z) = z/(z - \alpha)$$

i.e.

$$[h_d(n)] = 1, \alpha, \alpha^2, \alpha^3, \dots$$

then the all-zero filter (often called a *whitening filter*) represented by

$$A_d(z) = (z^{-1})(z - \alpha)$$

gives a resultant transfer function of

$$H_d(z) \cdot A_d(z) = 1$$

i.e. a unit sample. Fig 4.31 gives a realization which demonstrates the effect. The convolution of $a_d(n)$ with $h_d(n)$ should be zero everywhere except at $n = 0$, where it should be 1. It is thus possible to write out a set of N linear equations (where N is the number of poles in $H_d(z)$) which can be solved to give $a_d(n)$ directly. The important result, however, is that even when $H_d(z)$ is of order greater than N, and/or is not all-pole, it is nonetheless possible to find the impulse response set $a_d(n)$ which will minimize the mean squared error between the actual output of the cascade and the desired unit pulse at $n = 0$, with zero output for $n \neq 0$.

All-pole filters   The significance of the above result is that it gives a design method for making an Nth order all-pole filter approximate, in a least squared sense, an arbitrary transfer function. It is simply necessary to realize an all-pole (IIR) filter whose poles are the zeros

186

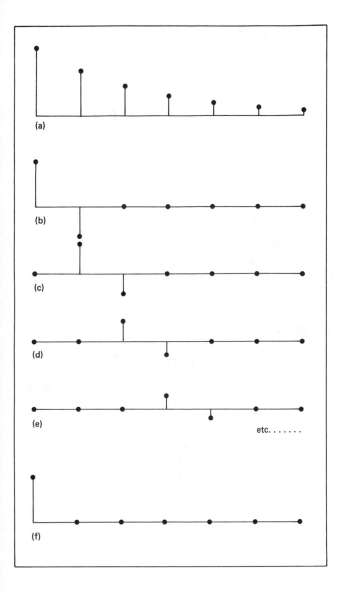

Fig 4.31 Illustration of the (all-zero) whitening filter corresponding to a first-order discrete all-pole filter:

a) All-pole filter pulse response

b) Inverse filter pulse response

c), *d*), *e*) Formation of the convolution sum

f) Output of the cascaded filters *a*) and *b*)

of the appropriate whitening filter. Obviously the quality of the match will depend on how closely the desired $H_d(z)$ can be modelled by an Nth order all-pole transfer function.

In practical designs based on this technique, the desired impulse response $[h_d(n)]$ is first truncated (windowed) to some finite length, from $n = 0$ to L say. Although this means that least squares approximation is now performed on the desired frequency response *after* convolution with the function $\text{sinc}(fLT_s)$, the windowing guarantees that the filter designed will be stable and minimum phase.

The next step is to calculate the aperiodic autocorrelation function of $h_d(n)$ as defined by

$$\phi(n) = \sum_{i=0}^{L-1-n} h_d(i) \cdot h_d(i+n)$$

If the design is to be Nth order, then only the autocorrelation coefficients for $n = 0$ to N need to be calculated. From $\phi(n)$ a set of numbers called the *reflection coefficients* can be derived by an iterative process which is described in section 5.6. The reflection coefficients contain the same information as the feedforward coefficients of the all-zero whitening filter for $H_d(z)$. Indeed, the feedforward coefficients can be derived from the reflection coefficients. However, there is a realization structure which makes use of the reflection coefficients directly. It is known as the *lattice structure*. A second property of the lattice structure is that the reflection coefficients can be used in either a feedforward or a feedback lattice to realize a filter with either poles or zeros in a given set of locations. This means that, once the reflection coefficients of the whitening filter $A_d(z)$ have been found, it is a simple matter to implement the approximation to $H_d(z)$ as a feedback lattice with the same reflection coefficients. Fig 4.32 illustrates the all-pole recursive lattice structure.

Fig 4.32 Lattice filter structure for the realization of an all-pole design using reflection coefficients

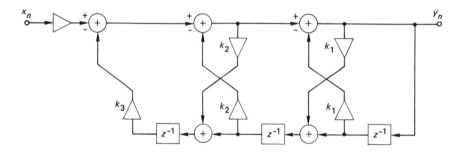

There are some additional advantages of the lattice structure although these are offset by the complication that twice as many multiplications are needed as in the standard all-pole IIR structure. First, all coefficients have a magnitude less than unity and, secondly, the summations are distributed evenly through the model, the combined effect of which is to make round-off errors less troublesome than in the standard IIR structure.

All-zero filters All-zero approximations to an arbitrary discrete transfer function can also be derived using the least squares method. $H_d(z)$ is first inverted, then the reflection coefficients of the whitening filter for the inverse transfer function are found. In this case the

188

whitening filter itself is the required design. An all-zero approximation will give a smaller maximum error when the desired frequency response has notches, and an all-pole approximation will be better when sharp edges must be realized. There is also a technique for combining the all-zero and all-pole design methods to yield an optimum approximation of given order which incorporates both poles and zeros. The realization structure in this case is a combined feedforward and feedback lattice. Reference 4.1 provides further information on this topic.

The least squares design techniques that have been described were initially developed to statistically model the spectra of random signals such as speech. Sometimes the nomenclature of statistical modelling is used to describe them. For example all-pole filters are referred to as *auto-regressive* or AR designs, and all-zero filters as *moving average* or MA designs. Filters with poles and zeros are ARMA designs. The feedforward coefficients in all-zero whitening filters are sometimes called the *linear prediction coefficients*.

The various IIR filter design techniques described above are summarised in Table 4.12.

Designs based on analog prototype

1 Bilinear Transform

$s = 2(z - 1)/T_s(z + 1)$

Most commonly used technique.
Use cascade filter structure.
Good frequency response match.
Poor impulse response match.

Predistortion:

$$f' = \frac{1}{\pi T_s} \tan(\pi f T_s)$$

2 Impulse Invariant

Use parallel filter structure.
Poor frequency response match (due to aliasing effects).
Good impulse response match at sampling points.
Relatively complex design procedure.

Designs for direct digital synthesis

3 Least Squares Design

Used for approximating an arbitrary transfer function. Either all-pole design (auto-regressive) or all-zero designs (moving average). Complex design process.

4 Direct Pole/Zero Placement

Suitable for simple resonator, notch and comb filters.
Quick, but limited.

Table 4.12 Summary of IIR filter design techniques

4.5 FIR filter design

4.5.1 Design by the window technique

When the desired frequency response $H_c(s)$ has a form which can be analytically inverse Fourier transformed (such as an ideal frequency selective filter, a differentiator or Hilbert transform filter), then it is possible to determine its corresponding impulse response. To make a realizable FIR filter, the appropriate impulse response must be multiplied by a window function which has finite width. It is then sampled at intervals of T_s to give the feedforward coefficients for the filter. The consequence of windowing is that the resulting frequency response of the filter is now the convolution of the wanted response (periodic in the frequency domain due to sampling) with the Fourier transform of the window.

Fig 4.33 illustrates the process for both rectangular and triangular windows applied to an ideal low-pass response. The large error in the actual realized response is mainly a consequence of the sharp transitions in the ideal response. When these are convolved with the sinc function transform of the rectangular window there are two effects. First the transition is forced to have a finite slope, which gets less steep as the window duration is reduced. Secondly the actual response oscillates in the pass and stopbands in the vicinity of the transition. The magnitude of the oscillation is constant and independent of the window length, an effect known as Gibbs phenomena, which is illustrated in Fig 4.34. (Note: the number of cycles per frequency span is affected by the window length.)

When shaped windows are used in place of the rectangular window, the trade-off between the slope of the transition and the oscillation magnitude (and hence the passband ripple and minimum stopband attenuation) can be varied. A large number of different window functions have been developed, the properties of some of which are tabulated in Table 4.13. In all cases the magnitude of the transition region and the attenuation of the first stopband lobe are given, for a desired ideal transition. It has been assumed that the ideal impulse response has been delayed by $MT_s/2$ seconds (where MT_s is the length of the window) to make the filter causal. The delayed ideal response is sampled at $t = nT_s$ for $n = 0$ to M and the samples are then multiplied by $W(n)$ to yield the required FIR filter feedforward coefficients.

The rectangular window has the steepest transition for a given window width but the worst sidelobe performance. Of particular interest is the Kaiser window which incorporates a variable parameter β giving continuous control over the transition width

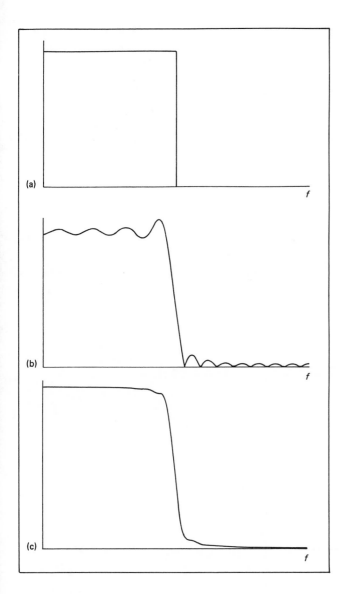

Fig 4.33 *a*) Ideal low-pass filter characteristic
b) Resulting characteristic when impulse response is truncated with a rectangular window
c) Resulting characteristic when impulse response is truncated with a double-width triangular window

Type	$\omega(n)$ $0 \leqslant n \leqslant M$	Transition width f_s/M	Max. stopband ripple (dB)
Rectangular	1	0.9	-21
Hanning	$\sin^2(\pi n/M)$	3.1	-44
Hamming	$0.54 - 0.46\cos(2\pi n/M)$	3.3	-53
Blackman		5.4	-74

Table 4.13 Realized response characteristics for various window functions, assuming wanted response is ideal transition

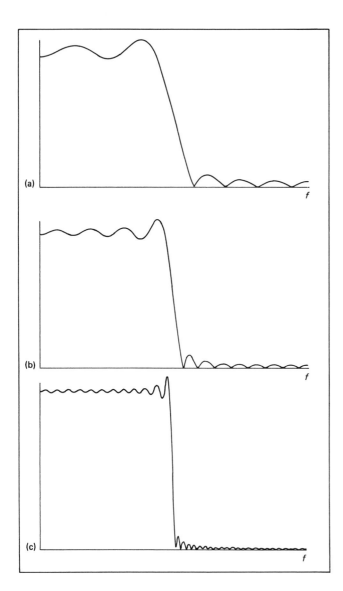

Fig 4.34 Illustration of Gibbs effect for rectangular window:

a) Window length = 16 samples

b) Window length = 32 samples

c) Window length = 64 samples

β	Transition width f_s/M	Max. stopband ripple (dB)
2.0	1.5	− 29
5.0	3.2	− 54
10.0	6.4	− 99

Table 4.14 Kaiser window characteristics

versus stopband ripple trade-off (cf. Table 4.14). The Kaiser window is defined as

$$W(n) = \frac{I_0[\beta\sqrt{(1 - (1 - 2n)/M^2)}]}{I_0[\beta]}$$

where

$$I_0[\beta] = 1 + \sum_{m=1}^{\infty} [(\beta/2)^m/m!]^2$$

(the first 15 terms are usually adequate).

The Kaiser window is well suited to designing FIR filters which have to comply with a specified mask. The ideal response is first defined, then the mask drawn in symmetrically around the pass band and transition regions, allowing some stopband ripple. The empirical formula below (Ref. 4.1) is used to decide the filter order M, where δ is the peak ripple amplitude. Note that, for filters designed using the window method, the passband and stopband ripple is the same, i.e. $\delta_p = \delta_s$ as previously defined.

$$\text{Filter order } M = \frac{(-20 \log \delta - 7.95)}{14.36 \text{ (transition width in Hz) } T_s}$$

$$\text{Passband ripple} = 20 \log_{10}[(1 + \delta)/(1 - \delta)] \text{ dB}$$

$$\text{Stopband attenuation} = -20 \log_{10}(\delta) \text{ dB}$$

Once the order has been decided, the value of β needed is derived from

$$\begin{array}{ll} \beta = 0.1102 \, (A - 8.7) & A \geqslant 50 \\ = 0.5842 \, (A - 21)^{0.4} & 21 \leqslant A \leqslant 50 \end{array}$$

where $A = -20 \log \delta$.

Because the formulae are empirical, the final design may not fully comply with the mask, in which case either β should be changed, or the order should be increased. (For high-pass filters the order should be odd, cf. section 4.3.)

Fig 4.35 gives an example design in which the mask corresponds to that used in the IIR examples shown in Figs 4.26 to 4.29. The FIR filter has an order of 27 (one less than the number of taps) compared with only 4 needed for the elliptic IIR design, but unlike the IIR design the phase response is linear.

4.5.2 Design by the equiripple technique

As was seen in the IIR case, the lowest-order filter which meets a given tolerance mask will be one with an equiripple passband and

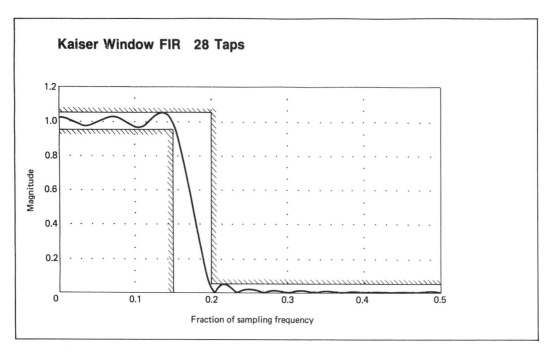

Fig 4.35 FIR filter design
using Kaiser window method

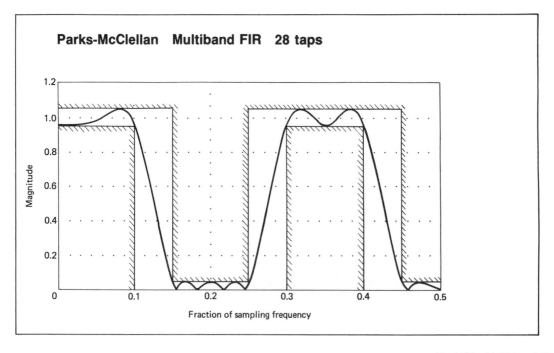

Fig 4.36 Multi-band filter
design using Parks-McClellan
algorithm

stop band. A technique for designing FIR filters on this basis was developed by T. W. Parks and J. H. McClellan (Ref. 4.2) using an iterative method based on the Remez exchange algorithm (Ref. 4.3).

When using this method, the transition width and filter order are defined and the filter coefficients adjusted iteratively until the response is equiripple. If desired, weighting factors can be applied to the pass and stopband regions of the response, resulting in different degrees of passband and stopband ripple. Fig 4.36 gives an example mask for a multi-band filter to be designed by the equiripple method. The method is inherently cut and try, since for a given filter order and transition width, the approximation error that results can only be found by doing the design. Fortunately empirical formulae have been published (Ref. 4.1) for estimating the order of filter needed to comply with a given mask. One simple formula is

$$\text{Filter order } M \approx \frac{-10 \log(\delta_P \delta_s) - 13}{14.6 \text{ (transition width in Hz) } T_s}$$

$$\text{Passband ripple} = 20 \log_{10}[(1 + \delta_p)/(1 - \delta_p)] \text{ dB}$$

$$\text{Stopband attenuation } R_{sb} = -20 \log_{10}(\delta_s) \text{ dB}$$

The main attraction of the equiripple method is the ability to separately tailor the passband and stopband responses. This flexibility in design comes at a price, namely the complexity and run time, of the design software. CAD packages are available for the IBM PC, but are relatively slow for large filter orders.

Fig 4.37 gives an example FIR filter designed using the equiripple method which complies with the mask used in the IIR examples, Figs 4.26 to 4.29. The filter order is now 23, compared with 27 using the Kaiser window design method.

4.5.3   Design by frequency sampling

Up to this point, most of the discussion has been concerned with the design of frequency selective filters. We now present a technique for synthesizing an approximation to an equivalent analog transfer function $H_c(s)$ that may be completely arbitrary, subject only to the constraint that the definition is restricted to the interval 0 to $f_s/2$. The magnitude and phase response may be independently specified over this range.

The desired (complex) $H_c(s)$ is first made periodic in the frequency domain and then sampled at intervals of $f_p = f_s/M$, where M is the order of the FIR filter to be designed. This gives rise to M complex samples overall. The inverse discrete Fourier transform IDFT of the set of frequency domain samples is now taken (cf. Chapter 5). The resulting coefficients $x(n)$ are then mapped onto the feedforward

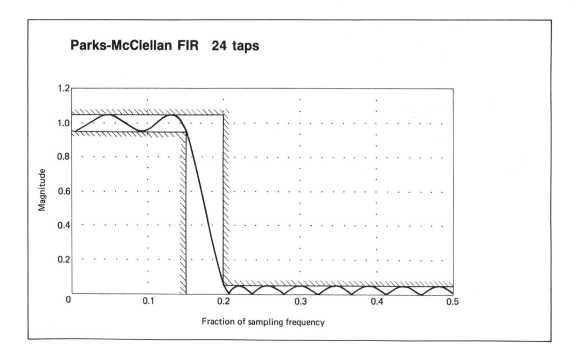

Parks-McClellan FIR   24 taps

coefficients of a causal version of the desired filter. The mapping for *M* even is

Fig 4.37   Low-pass FIR filter design using Parks-McClellan algorithm

$$
\begin{aligned}
h(n) &= x(M/2)/2 & n &= 0 \\
&= x(M/2 + n) & 1 &\leqslant n \leqslant M/2 - 1 \\
&= x(n - M/2) & M/2 &\leqslant n \leqslant M - 1 \\
&= x(M/2)/2 & n &= M
\end{aligned}
$$

For *M* odd, the frequency domain is sampled as for an $M + 1$ design, but before finding the inverse DFT, the sample corresponding to $f_s/2$ is replaced by $M + 2$ zero samples, so that the inverse DFT how has length $2M + 2$. The feedforward coefficients are derived from the resultant IDFT values as follows:

$$
\begin{aligned}
h(n) &= x(M)/2 & n &= 0 \\
&= x(M + 2n) & 1 &\leqslant 2n \leqslant M - 1 \\
&= x(2n - M) & M &\leqslant 2n \leqslant 2M - 1 \\
&= x(M)/2 & n &= M
\end{aligned}
$$

Once the feedforward coefficients have been found, it is possible to determine the *z*-plane zeros and hence determine the frequency and phase response of the filter produced. It will perform exactly as specified at frequencies corresponding to harmonics of the impulse response duration (MT_s) (apart from the extra linear phase shift term needed to make the filter causal), but will deviate in between these frequencies. The actual frequency response is the convolution of the

196

sampled, desired frequency response with sinc[fMT_s]. This is because the time domain equivalent of a sampled frequency response is in fact periodic, whereas in the filter implementation only one period of the impulse response is used.

We have seen that convolution with a sinc function causes ripple in the vicinity of sharp transitions. It is thus important that the transitions in the desired frequency response are sufficiently gradual to be accommodated with the order of filter used.

If the derived filter coefficients are subsequently multiplied by a non-rectangular window function, the ripple will be reduced. However the frequency response will no longer exactly meet the desired response at the harmonic frequencies.

4.5.4 Least squares design

The major disadvantage of FIR filters is the much larger order needed to approximate a given characteristic compared with an IIR filter. As a result FIR filters are used, in the main, only when linear phase is important. Much lower-order FIR filters can be realized when phase is not critical, using the least squares design technique, although the order is still larger than IIR equivalents. The design technique is described in section 4.4.5.

4.5.5 FFT-based filters

FIR filtering is effectively a convolution of the filter impulse response with the continuous input signal. In Chapter 5, section 5.6, a technique for fast convolution is introduced, based on the fast Fourier transform. When the order of the FIR filter is very large this technique may give a faster implementation than any of the above. The concept is explored further in Chapter 5.

Table 4.15 summarises the major properties of each of the FIR filter design techniques.

4.5.6 The differentiator and Hilbert transform filter

Two special linear phase transfer functions realizable as discrete FIR filters are the differentiator and Hilbert transform filter. The ideal transfer functions are

$$H(2\pi f) = j2\pi f \qquad \text{for the differentiator}$$

and

$$H(2\pi f) = -j \, \text{sgn}(f) \qquad \text{for the Hilbert transform}$$

where $\text{sgn}(f) = 1$ for $f > 0$ and -1 for $f < 0$.

1 *Window Method*	Allows tailoring of passband and stopband characteristics depending on choice of window. Less efficient than equiripple technique. Kaiser window offers good flexibility. Attenuation increases with distance from transition. Passband ripple δ_p = stopband ripple δ_s.

Approximate filter order $M = \dfrac{-20\log(\delta) - 7.95}{14.36fT_s}$

(f = transition width in Hz, $\delta = \delta_p = \delta_s$)

2 *Equiripple Method* (Parks–McClellan)	Optimal design procedure. Relatively long computation time.

Approximate filter order $M = \dfrac{-10\log(\delta_p\,\delta_s) - 13.0}{14.6fT_s}$

3 *Frequency Sampling*	Synthesis of arbitrary transfer function over range 0 to $f_s/2$.
4 *Least Squares Design*	Tends to give minimum filter length by sacrificing linear phase.
5 *FFT-based Filter Implementation*	Useful for realization of equalizer filters. Can be very efficient for long filter lengths.

Passband ripple = $20\log[(1 + \delta_p)/(1 - \delta_p)]$ dB

Stopband attenuation = $-20\log(\delta_s)$ dB

Table 4.15 Summary of FIR filter design techniques

The responses have $+\pi/2$ and $-\pi/2$ radians phase shift respectively for all positive frequencies but, when realized, there is always an added linear phase shift term to make the filter causal. Because discrete filters have periodic transfer functions it is only possible for $H_c(s)$ to approximate the ideal transfer function over, at most, the range $-f_s/2 \leqslant f \leqslant f_s/2$. However, since this implies a frequency response with finite duration in the frequency domain, an infinitely long FIR filter would be needed to realize either characteristic accurately over the full range.

We have already seen that imposing a finite duration limit on the impulse response smooths out the frequency domain transitions and introduces passband ripple, which is greatest near the transition. For the Hilbert transform filter there is a sharp transition at zero frequency, and others at $\pm f_s/2$ and every f_s thereafter (remember that the discrete transfer function is periodic). For the differentiator, transitions occur at the same places except that none occur at $f = 0$. To make allowance for the finite length of the filter, the mask should tolerate a symmetrical smooth change in the realized filter response in the region of the ideal response transitions, and allow some passband ripple in the magnitude characteristic. Fig 4.24 shows a suitable mask for each type of filter. Note that in the case of the differentiator, the passband ripple mask gets smaller as the distance from the transition increases. We know this is possible, since the ripple is worst nearest the transition. It is also desirable, so that the ripple is a fixed proportion of the nominal gain.

Note that, although the realized magnitude response is only an approximation to the ideal, *the phase response is exactly correct* (apart from the delay-induced linear phase variation). This condition is entirely controlled by the odd symmetry of the FIR filter coefficients.

If the mask is made symmetrical about $f_s/4$, the resulting FIR filter will have every second coefficient zero, reducing the implementation complexity by a factor of two.

As was pointed out in section 4.3, if the order of the realization is odd (an even number of taps), the mask passband may extend right up to $f_s/2$ (i.e. a high-pass filter response). The only disadvantage of an even tap filter design is that the delay is an odd number of half samples (cf. section 4.2.5). If, for example, the output of an even-tap Hilbert transform filter is to be compared with a delayed version of the input signal to achieve quadrature components, the input signal must in addition be passed through an even-tap linear phase FIR filter.

Both filter types are amenable to either window or equiripple design techniques. Figs 4.38 and 4.39 show typical realized responses and Fig 4.40 gives the unit pulse response and phase response corresponding to the design in Fig 4.38.

A Hilbert transform can also be realized by a pair of filters (Fig 4.41), the output of one being the Hilbert transform of the other. A very simple design procedure for FIR filter implementations is the SIN/COS transform of a low-pass filter prototype.

The coefficients $h(nT_s)$ of the low-pass filter are transformed into coefficients $h_1(nT_s)$ and $h_2(nT_s)$ of a pair of bandpass filters, using the following simple procedure:

$$h_1(nT_s) = 2h(nT_s) \cdot \sin(2\pi \cdot f_0 \cdot nT_s)$$

$$h_2(nT_s) = 2h(nT_s) \cdot \cos(2\pi \cdot f_0 \cdot nT_s)$$

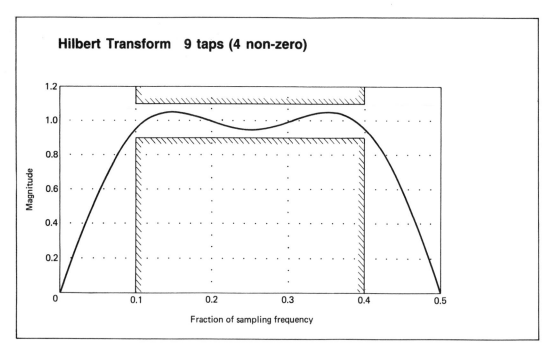

Fig 4.38 Hilbert transform filter design using a mask symmetrical about $f_s/4$ to force every second coefficient to zero

Fig 4.39 Hilbert transform filter with even number of taps (note: mask can extend to $f_s/2$)

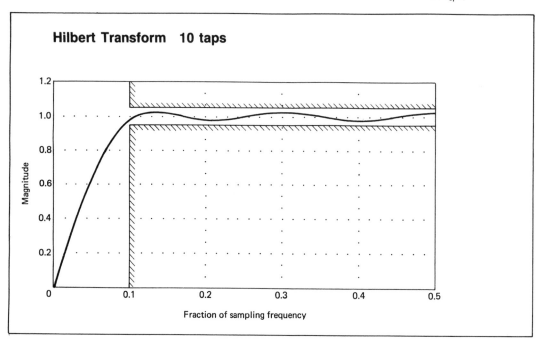

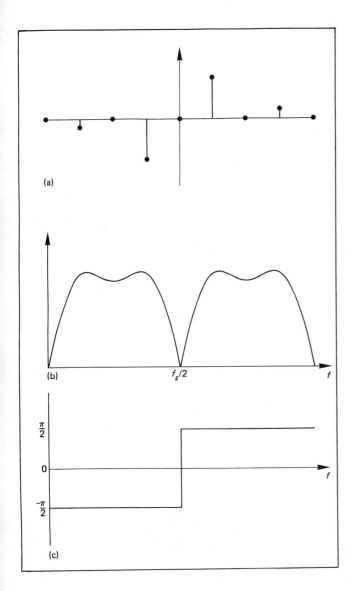

(a)

(b) $f_s/2$ f

(c) $\frac{\pi}{2}$ 0 $\frac{-\pi}{2}$ f

Fig 4.40 Hilbert transform response for filter in Fig 4.39:
- *a*) Impulse response
- *b*) Magnitude response
- *c*) Phase response

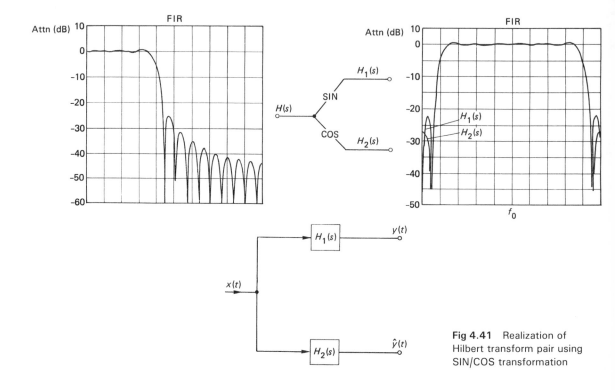

Fig 4.41 Realization of Hilbert transform pair using SIN/COS transformation

This results in a bandpass filter pair as illustrated in Fig 4.41, centred on a frequency f_0, and with a bandwidth equal to twice that of the low-pass filter prototype.

If f_0 is equal to $f_s/4$, then every other term in the SIN/COS transform goes to zero and consequently every alternate filter coefficient is zero. This is hardly surprising since the resulting bandpass filter response is symmetrical about a quarter of the sampling frequency.

4.6 Handling quantization and coefficient round-off effects

There are four possible sources of degradation in digital filters that are absent, or at least take a different form, in analog designs. They are:

Signal quantization
Coefficient quantization
Truncation noise
Internal overflow
Dynamic range constraints.

Signal quantization

Signal quantization, arising from the representation of a waveform sample by a finite number of bits, has been discussed in Chapter 1. When the output D/A converter resolution is less than the internal resolution of the processor, there will be an additional injection of quantization noise at the output. The two quantization noise components will add on a power basis. If, for example, 12-bit converters are used with 16-bit internal representation there will a contribution of noise from the input conversion, shaped by subsequent digital filtering, and a further white component due to the reduction from 16 to 12 bits at the output. For most applications, the level of noise attributable to this source is negligible.

Coefficient quantization

For filter coefficients stored as 16-bit numbers (Q15 format), the value of the coefficients that can be represented ranges from 0.999969 to -1.0 in increments of 0.000031. This quantization of coefficients means that the poles and zeros of the realized digital filter will be slightly offset from their correct positions. The consequences of pole/zero misalignment can be categorized as follows:

a) It becomes more significant, the tighter the specification being attempted.
b) It affects IIR filters more than FIR.
c) When 12 to 16 bits are used to represent the coefficients, little degradation is usually observed.

For FIR filters the effect is not serious, the filter remaining stable regardless of quantization error. Unit circle zeros remain on the unit circle, thus infinite attenuation notches are retained, although the notch frequency may be changed slightly. Most importantly, the linear phase characteristic (if present) will not be affected since the all-important symmetry of the coefficients is maintained even with quantization error.

In IIR filters the recursive nature of the implementation means that quantization errors in the feedback coefficients can push poles from just inside to just outside the unit circle, resulting in an unstable and thus unusable filter. Filters with poles close to the unit circle are thus most vulnerable in this respect. As explained earlier, one possible solution is to move the poles back from the unit circle by choosing a lower sampling rate—decimation. When this is not feasible, double precision arithmetic may be necessary.

The one IIR structure which is unconditionally stable is the lattice, since it can be proven that its poles lie entirely within the unit circle when the reflection coefficients are $\leqslant 1$. Of the other IIR structures, the cascade form is much more robust to quantization errors than the parallel form due to the isolation of error in a particular coefficient to a single pole pair. Also of relevance is the property that any unit circle zero pair associated with one of the cascaded sections will remain on the unit circle, thus maintaining notches.

It has been found (Ref. 4.1) that for cascaded IIR structures, poles in the region of ± 1 are most sensitive to coefficient quantization effects. If there is a problem then the so-called *normal form structure* should be used in the cascade of second-order sections. In the normal form, instead of filtering the input variable directly, the transfer function is expressed in *state space* form and the state variables are filtered. The normal form is illustrated in Fig 4.42 for the realization of a second-order section with poles at $a \pm jb$. The sensitivity to coefficient quantization is reduced because the coefficients use the real and imaginary parts of the poles directly.

The most sensitive (and thus rarely used) IIR filter structure is the direct form implementation.

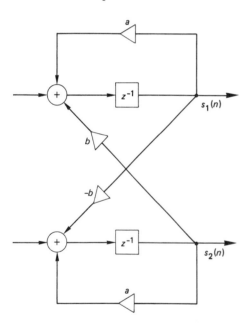

Fig 4.42 Normal-form realization of second-order state equations for system with poles at $a + jb$

Truncation noise

When multiplying one m-bit integer by a second m-bit integer, storage of the $2m$-bit product is required if full accuracy is to be preserved. For this reason, the accumulator in DSP devices is at least 'double width'. However when, as is the case with IIR filters, the products must be repeatedly stored then remultiplied, the least significant bits are repeatedly discarded since the memory locations are only single width. Each multiply/store operation therefore acts as a source of noise. The noise can be minimized by ensuring that the highest Q (narrowest bandwidth) poles are last in the cascade so that the bulk of the noise from previous sections is filtered by the last stage.

Internal overflow

In FIR filters the consequences of internal overflow are not serious; it is simply the case that one sample in the output is incorrect. The significance of this error is reduced if the processor is switched into saturation mode so that when overflow occurs the accumulator is loaded with the largest representable positive or negative number as the case may be, instead of changing sign (cf. section 3.6). Overflow can be entirely prevented in an FIR filter since the maximum output is known to be just the sum of the magnitudes of the coefficients —thus, prior to the filter, the input can be scaled down by the corresponding amount.

This approach however is unduly conservative. A more realistic approach is to evaluate the gain for a sinusoid at the peak gain frequency f_0. The input scaling factor S_{max} is then

$$S_{max} = 1/[H_d(z)]\,|_{z = \exp(j2\pi f_0)}$$

To prevent internal overflow in IIR filters, it is usually necessary to scale the signal at the input and between stages. To clarify this point, we return to the 5th order elliptic filter discussed in section 4.4.3. The required poles and zeros of each stage in the cascade are:

Stage 1: Pole at $z = 0.7228$
           Zero at $z = -1$
Stage 2: Poles at $z = 0.8135$ angle $\pm 26.54°$
           Zeros at $z = 1.0$ angle $\pm 76.76°$
Stage 3: Poles at $z = 0.9449$ angle $\pm 37.04°$
           Zeros at $z = 1.0$ angle $\pm 55.44°$

(Note: the pole nearest the unit circle is implemented in the last stage.)

The cascade of 3 stages can in fact by decomposed into 6, since each of the original stages can be represented as an all-pole filter between the input and the summing point and an all-zero filter between the summing point and the stage output.

Table 4.16 gives the individual gains of these all-pole and all-zero

	Stage 1 poles	Stage 1 zeros	Stage 2 poles	Stage 2 zeros	Stage 3 poles	Stage 3 zeros
f_1	3.607	2	4.849	1.541	2.601	0.8645
f_2	2.089	1.964	6.481	1.331	5.235	0.6547
f_3	1.647	1.896	4.146	1.138	15.48	0.4618

Table 4.16 Gains of each stage at three critical frequencies

sub-stages at three critical frequencies, which correspond to the resonant frequencies of each of the stages (when $2\pi f T_s$ is the same as the pole angle). It is assumed that unity scaling has been used throughout (i.e. a_0 in all stages is 1.0), and there is no interstage scaling.

If we assume a unit peak input signal, and calculate the signal levels at the various points in the cascade (Table 4.17), then for a fixed-point implementation, using Q15 format, it is immediately evident that overflow of the accumulator will occur. In order to ensure that no internal signal greater than unity will arise it is necessary to scale the input by 1/228.02.

	Input 1	Summing point 1	Input 2	Summing point 2	Input 3	Summing point 3	Output
f_1	1.0	3.607	7.214	34.98	53.9	140.2	121.2
f_2	1.0	2.089	4.065	26.34	35.06	183.57	120.18
f_3	1.0	1.67	3.12	12.49	14.73	228.02	105.32

Table 4.17   Signal levels at three critical frequencies, assuming no scaling

Stage 1 input scaler	Stage 1 output scaler	Stage 2 input scaler	Stage 2 output scaler	Stage 3 input scaler	Stage 3 output scaler
$\dfrac{1}{3.607}$	$\dfrac{1}{2}$	$\dfrac{1}{4.849}$	$\dfrac{1}{1.541}$	$\dfrac{1}{4.23}$	$\dfrac{1}{0.5314}$

Table 4.18   Optimum scaling factors

In fact it is better to distribute the scaling between the stages, reducing signals only as much as is necessary for the next sub-stage. This improves the signal-to-noise ratio. Table 4.18 gives the optimum scaling prior to each stage input, and also an output scale factor. The output scaler is necessary to ensure that the output of the all-zero filter does not exceed unity. Output scaling is achieved by multiplying all of the stage feedforward coefficients by the scaling factor. It would of course be possible to incorporate the input scaling for the following stage by this method, but this is undesirable since it will affect the filter characteristics by imposing coarser coefficient quantization.

Table 4.19 gives the signal levels in the filter, incorporating the scaling. The transfer functions and feedforward and feedback coefficients for the three stages are given below. The constant Ci represents the input scaling of each stage. The output scaling has been incorporated into the feedforward coefficients as discussed above. The figures in brackets give an alternative solution where all

	Stage 1 output	Scaler output	Stage 2 output	Scaler output	Stage 3 output	Scaler output
f_1	1.0	0.134	1.0	0.236	0.531	1.0
f_2	0.563	0.075	0.647	0.153	0.524	0.987
f_3	0.432	0.057	0.269	0.064	0.462	0.870

Table 4.19 Signal levels, scaling included

scale factors are a power of two and thus scaling can be achieved with appropriate right-shifting alone.

Stage 1

$$H_{d1}(z) = C1(z+1)/(z - 0.7228)$$

$C1 = 0.2772 \qquad (0.25)$

$a_{10} = 0.5 \qquad (0.5544)$
$a_{11} = 0.5 \qquad (0.5544)$

$b_{11} = 0.7228$

Stage 2

$$H_{d2}(z) = \frac{C2(z - 0.2290 - j0.9734)\ (z - 0.2290 + j0.9734)}{(z - 0.7278 - j0.3635)(z - 0.7278 + j0.3635)}$$

$C2 = 0.2062 \qquad (0.125)$

$a_{20} = 0.649 \qquad\quad (1.071)$
$a_{21} = -0.2972 \qquad (-0.4902)$
$a_{22} = 0.649 \qquad\quad (1.071)$

$b_{21} = 1.4556$
$b_{22} = -0.6618$

Note that in this section one of the feedback coefficients, b_{21}, is greater than unity. In the implementation, the signal sample would be first multiplied by 1.0 and then by the decimal part of the filter coefficient, with both products accumulated (remember the input has already been scaled down to prevent overflow).

Stage 3

$$H_{d3}(z) = \frac{C3(z - 0.5673 - j0.8235)\ (z - 0.5673 + j0.8235)}{(z - 0.7542 - j0.5692)(z - 0.7542 + j0.5692)}$$

$C3 = 0.2364 \qquad (0.125)$

$a_{30} = 1.882 \qquad\quad (3.559)$
$a_{31} = -2.1351 \qquad (4.037)$
$a_{32} = 1.882 \qquad\quad (3.559)$

$$b_{31} = 1.5084$$
$$b_{32} = -0.8928$$

The pertinent factors governing overall system performance are summarised in Table 4.20.

1	Coefficient Quantization	Source of noise and poor frequency response in filters. More prominent in IIR than FIR implementations. Increasingly important as transition width is reduced relative to f_s. 12–16 bits usually adequate for most IIR filters.
2	Truncation Noise	Reduced by sensible scaling between filter sections. Place highest Q pole sections last in a cascade structure.
3	Overflow	Common cause of assumed instability in IIR filters. Eliminated with careful gain scaling.

Table 4.20 Processor implementation considerations

4.7 Interpolation and decimation

Although not strictly a filtering operation, the processes of interpolation and decimation almost invariably involve filtering and are thus addressed in this chapter. It is one of the most powerful signal processing concepts, having a major bearing on system sampling rate, algorithm efficiency, anti-aliasing filter specifications and digital filter realizations.

Decimation is the term used to describe the ordered discarding of waveform samples, with the effect of reducing the effective sampling rate of the waveform. For example, if every other sample is discarded, the effective sampling rate is reduced by a factor of two. Provided that the new sampling rate still satisfies the Nyquist sampling criterion for the waveform, no aliasing will occur and the integrity of the digital representation remains intact.

What is the purpose of decimation? One significant application is that of *oversampling*. When sampling an analog waveform, it is essential that the sampling rate is greater than twice the maximum input signal frequency to avoid aliasing (unless aliasing is deliberately intended, cf. section 6.6). This criterion is usually met by placing an 'analog' anti-aliasing filter prior to the sampling circuitry to attenuate all components greater than $f_s/2$. Clearly, by adopting a sampling rate much greater than twice the maximum input

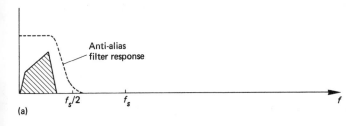

Fig 4.43 Effect of sampling rate on alias filter specification:
a) Low sampling rate
b) High sampling rate (oversampling)

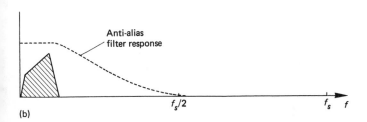

component frequency, the specification on the anti-aliasing filter becomes less stringent (Fig 4.43). In terms of the subsequent digital processing however, the sampling rate should be minimised, whilst still satisfying the Nyquist constraint, thereby maximising program execution time and minimising relative filter complexity. The solution to this apparent dilemma between internal and external sampling rates is to 'decimate' the deliberately 'oversampled' input signal to achieve the minimum sampling rate for internal processing.

The decimation process is illustrated graphically in Fig 4.44. Whilst it is possible to relax the specification of external, anti-aliasing filters, thereby achieving cheaper designs with better gain and phase characteristics, additional filtering now has to be performed digitally within the DSP as part of the decimation process. At each decimation stage, it is essential to ensure that the signal to be decimated occupies a bandwidth no greater than $\frac{1}{2} f_s/M$, where M is the sampling rate reduction factor (or at least contains no components that will be aliased into the band of interest). For the most common decimation factor of $M = 2$, this requires that the signal prior to decimation be subject to a low-pass filter that ideally attenuates all components greater than $f_s/4$. By making the filter transition region symmetrical about $f_s/4$, and using an FIR realization, alternate filter coefficients become zero (cf. section 4.5), and very efficient decimation filters can be implemented.

For large orders of decimation, it is usually more efficient to cascade an appropriate number of second-order stages than to implement one Mth order decimation, as the order and complexity of the decimation filter become excessive. By using symmetrical filtering techniques, the same decimation filter coefficients can be

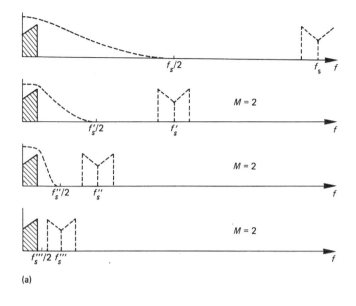

Fig 4.44 Decimation
process:
   a) Three stages of
decimation by two
   b) One stage of
decimation by eight

(a)

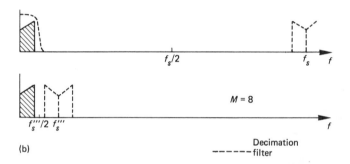

(b)

Decimation
— — — — filter

used for each of the cascaded second-order stages. These factors are
emphasised in the bandpass filter design example to follow.

    Whenever processing is to be performed on a waveform which
occupies a bandwidth much smaller than $f_s/2$, then decimation will
usually result in a more efficient algorithm. For bandpass signals, the
decimation process is often coupled with quadrature demodulation
to obtain low-pass equivalent waveforms. Quadrature demodulation
techniques are discussed in section 6.3.

    Having minimised the sampling frequency for optimum digital
processing efficiency, it is often desirable to increase the sampling
rate before D/A conversion to permit relaxation of the analog
reconstruction filter specification. The process is termed **interpolation**, and is achieved by interposing new samples between the
existing samples. As for decimation, interpolation almost always
involves subsequent filtering. There are two obvious choices of

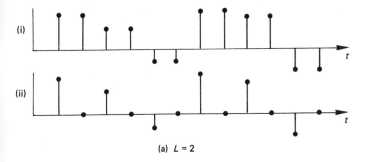

(i)

(ii)

(a) L = 2

Fig 4.45 Interpolation in
the time domain:
  i) Equal-sample insertion
  ii) Zero-sample insertion

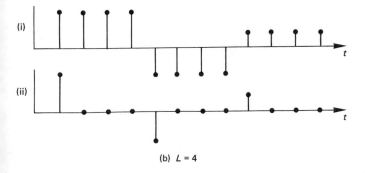

(i)

(ii)

(b) L = 4

sample values to insert: sample values equal to the most recent
existing sample, or sample values of zero. Both methods of interpolation are illustrated in Fig 4.45. It is most common to use
zero-value samples as this simplifies the digital filtering process
considerably. Equal-value samples on the other hand result in a
greater signal energy in the interpolated signal.

Simply inserting samples does not in itself achieve a true increase
in sampling rate since this would require the sample values to
correspond with those of the actual waveform at the new sampling
points. To achieve this result, the desired sample values are found by
interpolating between the original waveform samples, using a filter.
Fig 4.46 gives an illustration of interpolation in the frequency
domain, from which it can be seen that this filter, like that used for
decimation, must attenuate the 'alias' components about $f_s/2$, and
for simplicity should be an FIR filter with symmetry about $f_s/4$,
where f_s is the new sampling rate. A degree of interpolation L is
typically achieved using an appropriate number of second-order
stages as for decimation, although a single interpolation stage can be
entertained.

Sampling rate changes by a non-integer ratio are possible by
cascading a rate L interpolator with a rate M decimator. In this

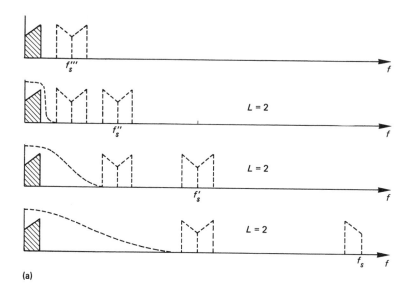

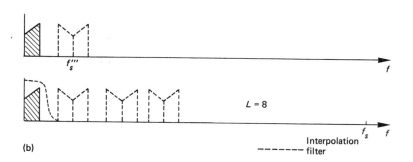

Fig 4.46 Interpolation process:
a) Three stages of interpolation by two
b) One stage of interpolation by eight

(a)

(b)

Interpolation
------- filter

configuration, the interpolator output filter can also be used as the first-stage decimation filter.

When interpolation is used to increase the sampling rate prior to outputting the signal, it is essential that the samples are fed to the D/A at *evenly spaced* time intervals within the main processing loop. If this is not the case, imperfect suppression of alias components will result.

Narrow bandpass filtering using decimation and interpolation
To further illustrate the potential of decimation and interpolation consider the design of a narrow, linear phase bandpass filter satisfying the mask specification as shown in Fig 4.47. This calls for a passband ripple of less than 0.15 dB and a stopband attenuation of greater than 50 dB. To implement this filter directly would require 231 taps assuming an FIR equiripple design technique—a time and

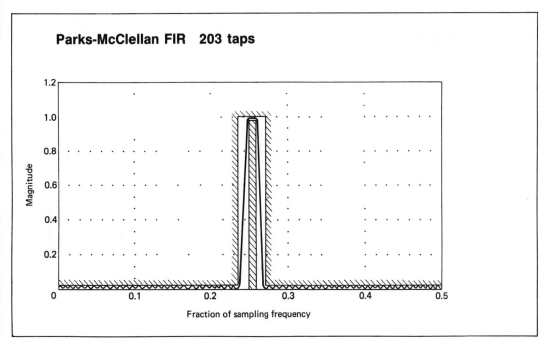

Parks-McClellan FIR   203 taps

Magnitude (y-axis): 1.2, 1.0, 0.8, 0.6, 0.4, 0.2

Fraction of sampling frequency (x-axis): 0, 0.1, 0.2, 0.3, 0.4, 0.5

Fig 4.47 Filter mask for narrow bandpass response

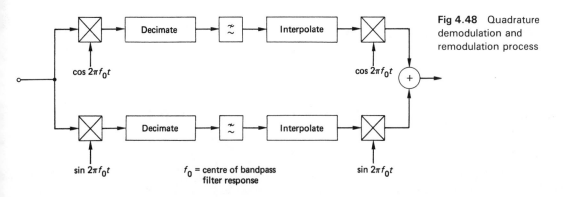

$\cos 2\pi f_0 t$

$\cos 2\pi f_0 t$

$+$

$\sin 2\pi f_0 t$

f_0 = centre of bandpass filter response

$\sin 2\pi f_0 t$

Decimate — Interpolate — Decimate — Interpolate

Fig 4.48 Quadrature demodulation and remodulation process

memory intensive task. It will become apparent that this filter can be realized far more efficiently using decimation and interpolation techniques.

In order to benefit from decimation, it is first necessary to convert the bandpass signal into its baseband equivalent. This involves quadrature demodulation about the centre of the bandpass signal spectrum as illustrated in Fig 4.48 (cf. 2.4.1 and 6.3). The process of decimation can now be used to reduce the sampling rate, in this example by a factor of 8. The decimation filters must satisfy the

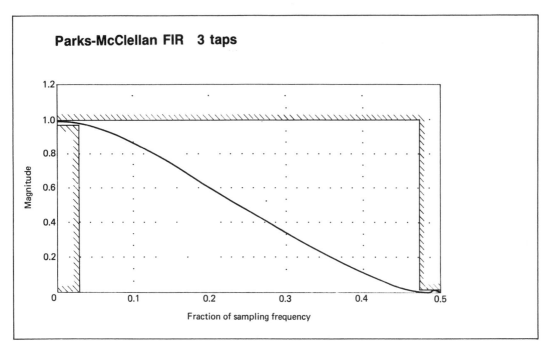

Fig 4.49 Mask for interpolation/decimation filter

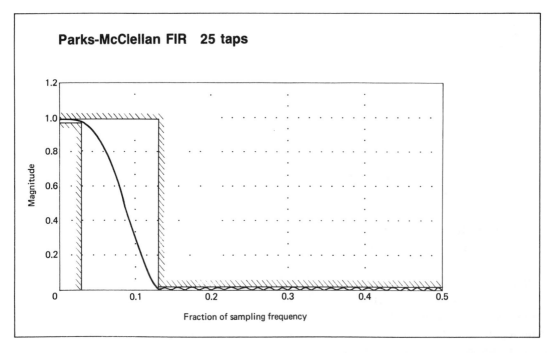

Fig 4.50 Mask for low-pass equivalent of bandpass filter

mask given in Fig 4.49, the most stringent specification arising from the last filter stage. In the example given, a 3-tap FIR filter can satisfy the mask for each decimation stage with identical filter coefficients.

The mask for the low-pass equivalent of the bandpass filter after decimation is given in Fig 4.50 and requires 25 taps. (This is also assuming on an equiripple design procedure.)

In order to restore the filtered signals to the original spectral band, the converse of the quadrature demodulation process must be applied. Before this is undertaken, however, the sampling rate must first be raised to the original value to avoid aliasing. By virtue of the symmetry of the decimation and interpolation process, the interpolation filters can be identical to those used for decimation, i.e. 3-tap FIR realizations. In all, three interpolation stages are required.

The total number of multiply-accumulate operations (effective taps) involved in the new bandpass filter realization is $3 + 3 + 3 + 25 + 3 + 3 + 3$ per filter chain, i.e. a total of 86 effective taps. The number of coefficients required for realization, however, is only $3 + 25 = 28$. Table 4.21 gives a comparison of the filter parameters for both methods of filter implementation. Clearly the decimation/interpolation approach is considerably more efficient.

The effective delay introduced by the bandpass filter is found by summing the delays of each individual filter in the process, *referenced to the original sampling rate*. For example, the delay introduced by an N-tap filter operating at $1/M$ times the original sampling rate is

Delay $= (MN - 1)/2$ cycles

In the above example, the total delay is thus

$$D_{total} = 1 + 2\tfrac{1}{2} + 5\tfrac{1}{2} + 99\tfrac{1}{2} + 5\tfrac{1}{2} + 2\tfrac{1}{2} + 1$$

$$= 117\tfrac{1}{2} \quad \text{cycles}$$

(Note: this does not equate to the delay expected from the number of taps for the equivalent bandpass filter.)

	Direct filter realization	Dec/Int filter realization
Effective no. of multiply/ accumulates	231	86
Data memory for coefficients	116	28
Delay	115	117½

Table 4.21 Comparison of direct and indirect bandpass filter realizations

If it is required to compensate for the filter delay in a second processing path, then the total filter delay can be made a whole number of samples by replacing either the first decimation filter or last interpolation filter with one having an even number of taps. By using a 4-tap interpolating filter in the last stage of the above example, the total delay now becomes 118 cycles.

A more detailed discussion of the design and application of decimation/interpolation processes is given in Refs. 4.4 and 4.6.

4.8 Specialist filter types

This section gives a necessarily brief introduction to the definition and implementation of specialist filter types found in many of the more advanced signal processing applications such as speech coding and modem design. The treatment is by no means exhaustive, and references are given to supplement the material where possible.

4.8.1 Tunable filters

A tunable filter is defined here as a filter whose gain response can be varied according to some suitable control signal, in a real time or pseudo real time sense. An example in the analog world is the switched capacitor filter, with frequency response determined by the clock rate. This technique can be carried over to the DSP environment by dynamically altering the sampling rate. This can be achieved by varying the program execution time where this does not conflict with other signal processing functions, or by interpolation/ decimation techniques.

A second trivial realization of semi-tunable filters is to modify the filter coefficients by accessing a data set held in (program) memory. This technique can only accommodate a predefined set of filter characteristics, but is nonetheless a valid and often-used process. This is not to be confused with adaptive filtering which is discussed below.

Where it is required to implement a filter with constant bandwidth, but variable centre frequency, such as a tuned bandpass selection filter, a frequency translation technique is usually optimal. Such a technique has already been outlined in section 4.7, whereby a narrow bandpass filter is implemented by two low-pass filters, sandwiched between a quadrature frequency translation system. Tunability of the design presented can be achieved by varying the frequency of the local reference (Fig 4.48). An alternative mechanism is to 'slide' the signal to be filtered past the 'fixed' filter, using one of the frequency translation techniques discussed in Chapter 6.

One further technique involves the fast Fourier transform and its corresponding inverse Fourier transform. The FFT of a signal will yield its frequency components, making filtering a simple task of discarding unwanted component samples prior to the inverse FFT. This process, however, only results in an approximation to the filtered waveform. The technique is computational intensive but has significant attraction for dynamic filtering of signals, where the bandwidth of the information to be selected is varying rapidly with time. A particular example is the filtering of reference signals subject to Doppler shift caused by variable platform velocity.

4.8.2   Filter transformations

High-pass from low-pass   A low-pass filter realized using DSP techniques, and thus exhibiting well defined and time invariant gain characteristics, can be transformed into a high-pass filter by simply subtracting the filtered output from the delayed input. For linear phase FIR filters, this process can be exact, since perfect delay matching is possible. For IIR filters, the performance depends on the group delay characteristics of the low-pass filter source. (The same technique is equally applicable to bandpass and bandstop transformations.) Examples of this transformation are given in Fig 4.51.

For an FIR filter, a low-pass to high-pass transformation can also be achieved simply by reversing the sign of alternate filter coefficients, i.e.

$$h_1(n) = (-1)^n \cdot h_2(n)$$

In this case, the transformed filter response is the mirror image of the original, mirrored about $f_s/2$ (Fig 4.52a). The same technique applies to bandpass and bandstop filters. An example of a bandpass mirror transformation is given in Fig 4.52b. This particular transform is widely used in sub-band coders, in the form of a 'quadrature mirror' filter bank. Sub-band coders are discussed later in this section.

Bandpass from low-pass   As has been demonstrated in connection with the Hilbert transform filter (section 4.5.6), a bandpass filter can be derived from a low-pass prototype by using a simple SIN or COS transformation. This involves multiplying the filter coefficients (FIR filters only) by either

$$2\cos(2\pi f_0 n T_s) \quad \text{or} \quad 2\sin(2\pi f_0 n T_s)$$

where f_0 is the centre frequency of the desired bandpass response,

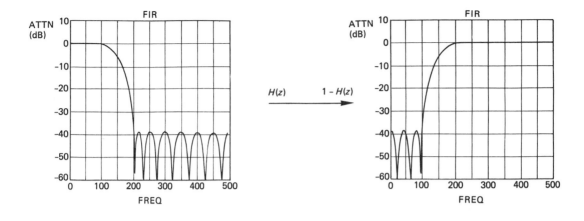

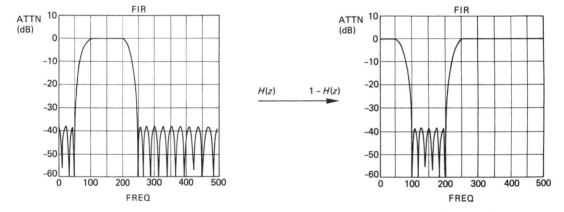

Fig 4.51 Complementary
low-pass/high-pass and
bandpass/bandstop filter
transformation

having a width twice that of the low-pass prototype. This technique
is particularly useful in data modem design for realizing matched
filters with a well defined low-pass characteristic, e.g. raised cosine,
in a bandpass form. Using both COS and SIN transforms together
gives rise to a Hilbert transform filter pair.

4.8.3 Mth-band filters

The property of a so-called Mth band filter is that one out of every M
samples in the impulse response is zero. In other words, for linear
phase FIR filters, every Mth coefficient is zero. The most obvious
example of this is the half-band filter ($M = 2$) which we have already

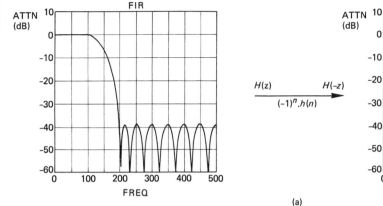
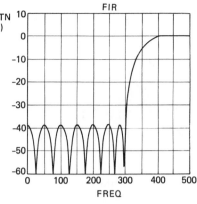

(a)

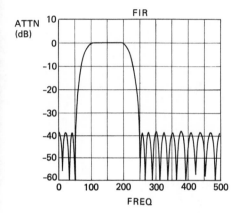
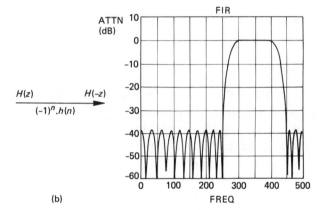

(b)

Fig 4.52 Mirror low-pass/high-pass and bandpass/bandpass filter transformations

encountered, formed by making the filter gain response symmetrical about $f_s/4$; here, every other filter coefficient is zero.

To realize an Mth-band filter, the filter transition region(s) must be symmetrical about $f_s/2M$, as one might expect, and the passband and stopband ripple satisfy the following condition [Refs. 4.7 and 4.8]:

$$\delta_p \leqslant (M-1)\delta_s$$

Mth-band filters, of order greater than two, generally find application as Mth degree decimation and interpolation filters. Design of Mth band filters $(M > 2)$ which meet the above constraints is not straightforward. The Parks-McClellan program can be used; however the design is sub-optimal [Ref. 4.8].

4.8.4   Sub-band filters

Sub-band filters are a type of filter particularly suited to the segmentation of a frequency band into a number of sub-bands, usually of equal width, and usually in such a way that the original band can be reconstructed without distortion, either in amplitude or phase. The driving force behind the development of sub-band filters has been speech coding systems, whereby each sub-band is coded separately, transmitted, and after decoding, recombined to form the original speech spectrum. The need for this process to be distortion-free is clearly apparent.

Quadrature mirror filters   One particular class of sub-band filters which has dominated this area is the quadrature mirror filter. The quadrature mirror filter structure is shown in Fig 4.53 and incorporates both transmit (analysis) filtering and receive (synthesis) filtering. The transfer functions of the various filters are such that the combined gain response is unity and the two transmit filters are *power complementary*, i.e.

$$| H_0(e^{j\omega}) |^2 + | H_1(e^{j\omega}) |^2 = 1$$

In the configuration given, H_0 and H_1 are a low-pass and high-pass mirror filter pair. (It is also possible to implement a quadrature mirror filter using quadrature processing techniques. In this case, the filters have an identical low-pass response, Ref. 4.9.)

Having divided the input spectrum into two sub-bands it is sensible to decimate the two bands to obtain the minimum sampling rate, before encoding. A similar interpolation stage is therefore needed in the receiver prior to recombination. The conventional quadrature mirror filter bank is thus configured as in Fig 4.54, with frequency mapping occurring as shown. To subdivide the spectrum further, quadrature mirror filters can be cascaded in a tree structure (Fig 4.55), each stage increasing the number of sub-bands by a factor of two.

The design of filters for the quadrature filter bank is covered in Ref. 4.9.

4.8.5   Adaptive filters

Adaptive filters, as the name suggests, are filters whose gain and/or phase response can be adapted with time. The most common application of adaptive filters is in channel equalizers for communications systems, to compensate for time-varying gain and phase distortion on a channel. In a fixed line telephone channel, the adaptation may be performed only periodically as the variations in

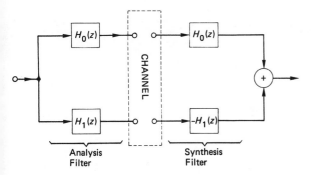

Fig 4.53 Quadrature mirror filter bank

Fig 4.54 Quadrature mirror filter bank realization using interpolation/decimation

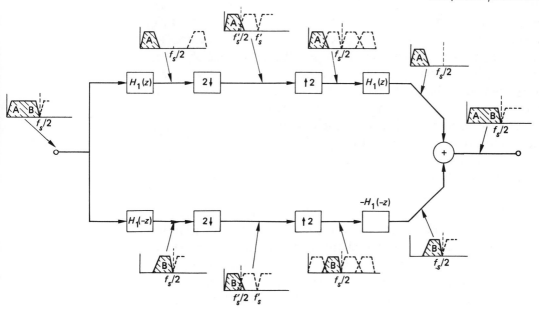

221

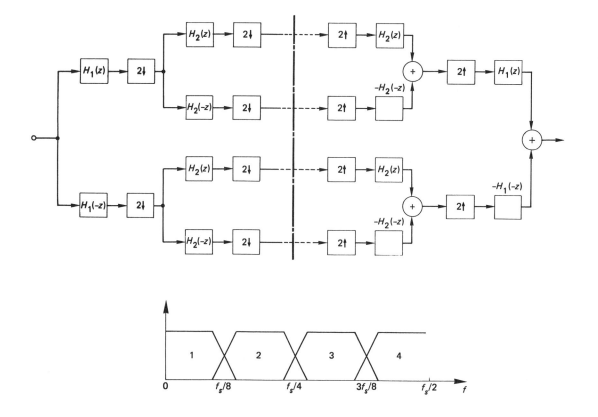

Fig 4.55 Tree structure for a 4-sub-band QMF bank

channel response are pseudo stationary in nature. With a mobile radio channel on the other hand, adaptation must be continuous, as the channel characteristics change dynamically with mobile position.

The most common form of adaptive filter is based on an FIR structure, whereby the coefficients (tap weights) are updated to realize the new gain and phase response. We have already seen (section 4.5) that such an FIR filter can be use to approximate an arbitrary gain and phase characteristic, the accuracy of the match being governed primarily by the order of the FIR filter used.

Probably the most difficult task in the design of adaptive filters is the algorithm used to determine the coefficients. The first difficulty lies in obtaining a measure of the gain and phase distortion of the channel. The second problem is how to make use of the information to adapt the filter (equalizer). There are in fact several techniques for achieving both operations, many of which are treated comprehensively in Ref. 4.10. One conceptually simple approach is to sound the channel using a series of equally spaced tones, the received magnitude and phase of which form the basis of a sampled frequency response of the channel. By computing the inverse of this response, and then performing the inverse discrete Fourier transform on the frequency samples, the sampled impulse response of the channel equalizer is obtained, yielding directly the FIR filter coefficients as outlined in section 4.5. The example is rather trivial in that no feedback control path is employed and the adaptation is discrete, requiring the sounding sequence to be transmitted before each coefficient update. The more advanced equalizer filters are designed to operate continuously, deriving a control (error) signal from the incoming information stream itself.

Detailed information on adaptive filter algorithms can be found in Ref. 4.10, and in the open literature. (Note: IIR filters can also be used in an adaptive sense, giving in most cases a more efficient implementation, but requiring considerably greater sophistication in the control algorithms.) Needless to say, the advent of DSP has opened up the whole field of adaptive filtering in commercial applications.

References

4.1   Jackson, *Digital Filters and Signal Processing*, Kluwer (1985)
4.2   McClellan, J. H., Parks T. W. and Rabiner, L. R., A computer program for designing optimum FIR linear phase digital filters, *IEEE Trans. AU–21,506* (1973).
4.3   Remez E., 'General Computational Methods of Tchebycheff Approximation', Kiev (Atomic Energy Translation 4491), pp. 1–85 (1957)
4.4   Crochiere, R. and Rabiner, L. R., Interpolation and decimation of digital signals—a tutorial review, *Proc. IEEE*, Vol. 69, 300 (1981)
4.5   *Computer Programs for Digital Filter Design*, IEEE Press (1976).
4.6   Crochiere, R. E. and Rabiner, L. R., *Multirate Digital Signal Processing*, Prentice-Hall (1983).
4.7   Vaidyanathan, P. P. and Nguyen, T. Q., A TRICK for the design of FIR half-band filters, *IEEE Trans. CAS–34*, pp. 292–296 (1987).
4.8   Mintzer, F., On half-band, third-band and Nth-band FIR filters and their design, *IEEE Trans. ASSP–30*, pp. 734–738 (1982).
4.9   Galand, C. R. and Nussbaumer, H. J., New quadrature mirror filter structures, *IEEE Trans. ASSP–32*, pp. 522–531 (1984).
4.10 Cowan, C. F. N. and Grant, P. M., *Adaptive Filters*, Prentice-Hall (1985).

5 Spectral analysis and the FFT

5.1 Introduction

Fourier transform based signal processing, which is often simply termed 'spectral analysis', is used extensively for modern-day engineering tasks, including speech recognition, vibration analysis, and biomedical engineering. Characteristic features which are entirely obscured in the time domain often become explicit in the frequency domain. A common example is the interference from domestic power lines, which, though masked in the time domain, appears clearly as discrete harmonics of 50–60 Hz in the frequency domain.

In recent years, a number of DSP-based products for audio spectral analysis have been launched, at the heart of which lies a frequency domain analysis technique known as the *Discrete Fourier Transform* (DFT), and, in particular, the *Fast Fourier Transform* (FFT) derivative. Implementation of the FFT in a fast and cost-effective manner has only really been made possible by the advent of current-generation DSP devices.

This chapter details the implementation of the DFT and FFT using DSP techniques and discusses some of the many novel applications of discrete spectral analysis in waveform processing. The first few sections seek to provide an understanding of the DFT and FFT process and its realization. The latter half of the chapter focuses on the practical application of the DFT to derive the spectral properties of a continuous signal from its sampled version. Because of the intimate relationship between Fourier transformation and convolution and correlation, it is often more efficient to implement these operations using FFT-based algorithms than conventional time domain techniques. A section is also devoted to this topic.

5.2 The Discrete Fourier Transform

In essence, the DFT is simply a mapping of one ordered set of N complex numbers to a different ordered set—the former conveying time domain information, the latter frequency domain information. The precise definition of the DFT is

$$X(k) = \sum_{n=0}^{N-1} x(n) \cdot \exp(-2\pi nk/N)$$

where n is used as the sequence member index (sample number) in the input discrete signal and k as the index for the transformed signal. At this stage we will concentrate on algorithms for implementing this definition—the interpretation will be deferred to a later section. There is also an inverse DFT (IDFT) which has the same form as the forward transform but with the argument of the exponential positive. A DFT processor may be used to calculate the IDFT by simply conjugating the input samples, and then conjugating the result.

The DFT expression can be expressed in matrix form as

$$
\begin{bmatrix}
X(0) \\
X(1) \\
\vdots \\
X(k) \\
X(N-1)
\end{bmatrix}
=
\begin{bmatrix}
W(0,0) & W(0,1) & \dots & W(0,n) & \dots & W(0,N-1) \\
W(1,0) & W(1,1) & \dots & W(1,n) & \dots & W(1,N-1) \\
\vdots & \vdots & & \vdots & & \vdots \\
W(k,0) & W(k,1) & \dots & W(k,n) & \dots & W(k,N-1) \\
W(N-1,0) & W(N-1,1) & \dots & W(N-1,n) & \dots & W(N-1,N-1)
\end{bmatrix}
\begin{bmatrix}
x(0) \\
x(1) \\
\vdots \\
x(n) \\
x(N-1)
\end{bmatrix}
$$

where n and k are used for the matrix column and row indices respectively. $W(k,n)$ is termed the **twiddle factor** and is usually denoted as

$$W(k,\ n) = W_N^i \qquad (i = nk)$$

It is a unit-magnitude complex number whose real and imaginary parts are the cosine and sine of the angle $(-2\pi i/N)$ radians. Thus,

$$W_N^i = \exp(-j2\pi i/N) = \cos(2\pi i/N) - j\,\sin(2\pi i/N)$$

We will use the terms 'twiddle superscript' and 'twiddle subscript' to refer to i and N respectively. Where there is no confusion as to the subscript, the notation for W_N^i will be simplified to Wi. Note that as i increases, the angle of Wi increases uniformly from zero in steps of $2\pi/N$ radians (Table 5.1). The coefficients in the first row and column of the matrix are thus all unity. In the computation of the DFT it is normal to recall the precalculated twiddle factors from a designated memory block/array.

Even though the maximum value of i ($= nk$) is $(N-1)^2$, only N twiddle factors need to be stored, since the cos and sin terms repeat for multiples of 2π, i.e.

$$W_N^i = W_N^{i \bmod N}$$

In Example 5.1 a Pascal implementation of the DFT is given. The input complex samples are stored in the real array XIN, with real and

i	Re(Wi)	Im(Wi)
0	1	0
1	0.707	− 0.707
2	0	− 1
3	− 0.707	− 0.707
4	− 1	0
5	− 0.707	0.707
6	0	1
7	0.707	0.707

Table 5.1 Twiddle factors for $N = 8$

imaginary components occupying consecutive positions. The output is stored similarly in XOUT. The output is derived by multiplication of the input array with the twiddle factor matrix—an $N \times N$ square matrix with complex coefficients stored in array **W**. The matrix coefficients are extracted from **W** using a predefined function, mod (N, i), which returns the value i mod N.

In Example 5.2, an equivalent DSP implementation of the DFT is given. Data page zero is used for input data, and the real and imaginary parts of the DFT are output one by one through port PAO.

Since the real and imaginary parts of the twiddle factor matrix, Wi, are $\cos(2\pi i/N)$ and $-\sin(2\pi i/N)$ respectively, the cos terms can be derived from a single table of sine values by offsetting the index by a quarter of a period.

In general, an N-point DFT requires N^2 complex multiplications, although some of these are multiplications by 1. In the somewhat trivial case of $N = 2$ and $N = 4$ the **W** matrices are

$$\begin{bmatrix} 1 & 1 \\ 1 & -1 \end{bmatrix} \text{ and } \begin{bmatrix} 1 & 1 & 1 & 1 \\ 1 & -j & -1 & j \\ 1 & -1 & 1 & -1 \\ 1 & j & -1 & -j \end{bmatrix} \text{ respectively}$$

and the entire DFT can be implemented without multiplications.

EXAMPLE No. 5.1

Pascal DFT.

```
begin
        for k := 0 to N-1 do begin
            xout[2*k] := 0;
            xout[2*k+1] := 0;
            for n := 0 to N-1 do begin
                i := mod(N,n*k);
                xout[2*k] := w[2*i] * xin[2*n] - w[2*i+1] * xin[2*n+1] + xout[2*k];
                xout[2*k+1] := w[2*i+1] * xin[2*n] + w[2*i] + xi[2*n+1] + xout[2*k+1];
            end;
        end;
    end.
```

227

TMS EXAMPLE No. 5.2

Example of a DFT, using a TMS 32010.

Eight point DFT on data stored in data page 0, in form real followed by imaginary in successive locations. Maintains 32 bit running sum. Results output sequentially on port PA0.

IDT	'DFT'		:program title
AORG	0		:program destination address
B	INIT		:send resets to initialisation

Define symbols for referencing constants and data memory locations.

N	EQU	8	:size of transform
PG1	EQU	128	:start address of data page 1
ONE	EQU	0	:to hold 1
NSTORE	EQU	1	:to hold N
QUARTN	EQU	2	:to hold N/4
K	EQU	3	:count for DFT point
MODMSK	EQU	4	:modulo mask for pointer
NCON	EQU	5	:number of values to initialise
I	EQU	5	:sine table pointer
SIN	EQU	6	:imaginary part of twiddle
COS	EQU	7	:real part of twiddle
SUMREH	EQU	8	:current real DFT sum (high)
SUMREL	EQU	9	:current real DFT sum (low)
SUMIMH	EQU	10	:current imaginary DFT sum (high)
SUMIML	EQU	11	:current imaginary DFT sum (low)

Place data table in program memory for initialising data memory.

TABLE	DATA	1	:for ONE
	DATA	N	:for NSTORE
	DATA	N/4	:for QUARTN
	DATA	0	:for K (row index zero on entry)
	DATA	N-1	:for modulo mask

Place lookup table of sine and cosine values in program memory.

SINE	DATA	0,23169
COSINE	DATA	32767,23169
	DATA	0,-23169
	DATA	-32767,-23169
	DATA	0,23169

Initialisation routine.

INIT	DINT	:interrupts disabled

```
            SOVM                        :overflow on
            LDPK    1                   :use data page 1

            LACK    TABLE               :source for data transfer
            LARK    AR0,ONE+PG1         :destination
            LARK    AR1,NCON-1          :number of values

RLOOP       LARP    AR0                 :ARP to AR0
            TBLR    *+,AR1              :transfer and update destination
            ADD     ONE                 :update source
            BANZ    RLOOP               :loop for rest
```

Outer loop initialisation.

```
KLOOP       LARP    AR1                 :ARP to AR1
            LARK    AR0,N-2             :to countdown rows
            LARK    AR1,0               :pointer to input data

            LAC     *+,15               :point to XR[0]
            SACH    SUMREH              :initialise running sums
            SACL    SUMREL

            LAC     *+,15               :point to XI[0]
            SACH    SUMIMH
            SACL    SUMIML

            ZAC
            SACL    I                   :set twiddle index to 0
```

Inner loop initialisation.

```
NLOOP       LARP    AR1
            LAC     I                   :load pointer
            ADD     K                   :update it
            AND     MODMSK              :with modulo wrap
            SACL    I                   :save new value

            LACK    TABLE               :point to table
            ADD     I                   :add offset pointer
            TBLR    SIN                 :read sine
            ADD     QUARTN              :adjust for cosine
            TBLR    COS                 :read cosine
```

DFT calculation.

```
            ZALH    SUMREH
            ADDS    SUMREL
            LT      COS
            MPY     *+
            LTA     SIN
            MPY     *-
            APAC
            SACH    SUMREH              :sumr = sumr + XR.cos + XI.sin
            SACL    SUMREL
```

```
            ZALH   SUMIMH
            ADDS   SUMIML
            MPY    *+
            SPAC
            LT     COS
            MPY    *+,AR0
            APAC
            SACH   SUMIMH        :sumi = sumi + XI.cos + XR.sin
            SACL   SUMIML

            BANZ   NLOOP         :repeat for n = 2 to N-1
```

Output results for current row.

```
            OUT    SUMREH,PA0
            OUT    SUMIMH,PA0

            LAC    K
            ADD    ONE
            SACL   K             :repeat for k = 1 to N-1

            SUB    NSTORE
            BLZ    KLOOP
STOP        B      STOP          :finished

            END
```

5.3   The Fast Fourier Transform

The Fast Fourier Transform, developed originally by Cooley and Tukey in 1965, is a significantly less computationally intensive method of evaluating the DFT, and thus particularly attractive for 'real time' spectral analysis using DSP technology. There are a number of types of FFT algorithm, but all share the common constraint of only working for certain values of N (the number of input samples). If this constraint is unacceptable, the full-blown DFT must be implemented.

The FFT reduces the number of complex multiplications involved from N^2 for the DFT to

$$N \log_2 (N)$$

For a 64-point FFT, this represents a reduction by a factor of 10.

5.3.1   Decomposition of large transforms

The FFT algorithm is based on the principle of computing a large transform via a number of smaller, more manageable transforms. As an example, consider the decomposition of an N-point transform, where N is even, into a pair of $N/2$-point transforms.

There are in fact two methods to achieve this result. The first, which is called *decimation-in-time* (DIT), implements the two $N/2$-point transforms using the even and odd elements of the input sequence respectively. These two transforms are then merged by a further $N/2$ two-point DFT to generate the desired output elements. The second method, called *decimation-in-frequency* (DIF), performs the same two sets of operations, but in reverse order. The input samples are firstly processed in pairs by $N/2$ two-point transforms, followed by the two $N/2$-point transforms which generate odd and even components of the output sequence directly.

Both methods are *in-place*, that is the same memory locations used for storing the DFT inputs can be overwritten, first with intermediate results, and then with the output results. No additional temporary storage is needed. The process is best illustrated by arranging the storage locations vertically and indicating the various processing stages horizontally, the consequences of each stage being the replacement of the original stored data with a new set. The diagram can be further simplified if the symbol

$$
\begin{array}{ll}
x(0) \diagdown \diagup X(0) & = x(0) + x(1) \\
x(1) \diagup \diagdown X(1) & = x(0) - x(1)
\end{array}
$$

is used to represent a two-point DFT.

Figs 5.1 and 5.2 illustrate the DIT and DIF decomposition algorithms. They show how the decomposition is carried out but give no insight as to why. The derivation is not difficult (see for example Ref. 5.1.) but beyond the scope of this book.

One way of remembering which element combinations (samples) are used in the first stage of the DIF decomposition is to arrange the element indices in columns of $N/2$ elements, filling the columns successively from the top. Thus for an 8-point transform we have for the DIF decomposition:

```
0 4
1 5
2 6
3 7
```

indicating that the four two-point DFTs have the pairs of elements indicated by the rows, i.e. 0 and 4, 1 and 5, 2 and 6, and 3 and 7. For the DIT algorithm the element indices are laid out in columns of 2;

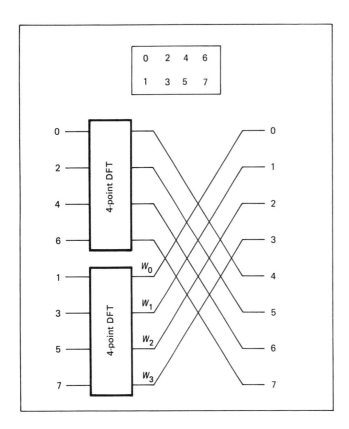

thus the first stage consists of two 4-point transforms with indices given by the rows of

0 2 4 6
1 3 5 7

The columns in this case identify the indices of the intermediate result elements that are combined in the second-stage DFT computations.

Two further features should be noted from the examples given. The first is that the element index associated with a particular storage location changes during the course of the processing. In the DIT algorithm the input is out of sequence (scrambled) and the output is in correct order. In the DIF the reverse is true. The second feature is that, following the first-stage transforms, some of the intermediate data elements are multiplied by a constant (complex) twiddle factor Wi.

Normally the two-point DFT and twiddle multiplication are combined in a single processing operation, which is termed a **butterfly**. In the DIT butterfly the input data is 'twiddled' before the DFT. In the DIF butterfly, the data is 'twiddled' after the butterfly.

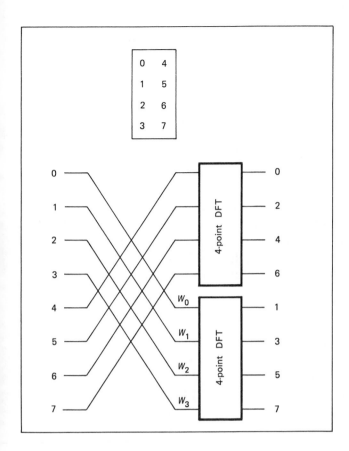

Fig 5.2 Decomposition of an 8-point DFT into two 4-point transforms using the decimation in frequency (DIF) algorithm

In both cases there are two complex multiplications and two complex additions. (The execution time of a butterfly is a standard benchmark performance indicator for signal processing hardware.)

In the next section the butterfly is examined in more detail. We will then be in a position to present one version of the FFT. It is a straightforward application of the decomposition principle just outlined, where the N-point transform is composed entirely of two-point transforms. The speed saving comes about because the time taken to perform two $(N-1)$-point transforms is less than the time taken to perform one N-point transform. Thus each decomposition results in a further speed increase.

The decomposition technique has a wider practical application when it is desired to perform the FFT on a block of data that is too large to fit into the available data memory space. The block is divided into two or more sub-blocks which can be accommodated, and each is transformed separately using the FFT. The pairs of data elements are then processed with a butterfly calculation to sequentially yield the full-size output block. The technique is illustrated in section 5.4.

5.3.2   Implementing radix-2 butterflies

As we have seen, at the heart of any FFT computation is an elementary processing operation known as the butterfly. The radix-2 butterfly is defined as having two inputs and two outputs. Higher-radix butterflies are used in certain circumstances to reduce the FFT computation time and these will be considered in later sections; here we focus on the implementation of the radix-2 butterfly.

Fig 5.3 gives three pictorial representations of the radix-2 DIF butterfly. One representation highlights the embedded two-point DFT, one uses signal flow graph symbols, and the third gives the conventional butterfly symbol. If the complex signal samples at the butterfly input are designated (XPR + jXPI) and (XQR + jXQI) respectively, and if WR and WI are used to represent the real and imaginary part of the twiddle factor, then the butterfly outputs are:

$$XP(out) = (XPR + XQR) + j(XPI + XQI)$$

$$XQ(out) = [(XPR - XQR) \cdot WR - (XPI - XQI) \cdot WI]$$
$$+ j[(XPI - XQI) \cdot WR + (XPR - XQR) \cdot WI]$$

Example 5.3 gives a Pascal procedure for implementing the radix-2 butterfly. The entire array of input sample values is passed to the procedure, along with the indices, designated *p* and *q*, of the two samples to be processed by the butterfly. Also passed are *wr* and *wi*, the real and imaginary parts of the twiddle factor. As for the previous example, the input signal array is real, with real and imaginary input elements held in consecutive array locations. Two temporary variables *ti* and *tr* are used internally. On completion, the *p* and *q* elements of the input array are overwritten with the results.

Example 5.4 gives a DSP realization of the DIF butterfly, modelled exactly on the Pascal procedure in Example 5.3, and using the same variable names. In this case, XPR, XQR, XPI, XQI, WI, WR, TI and TR are addresses in the current page of data RAM at which the associated variables are stored. The butterfly requires 23 machine cycles to execute.

The code in Example 5.4 has no protection against overflow. Assuming Q15 representation of the data, the maximum input sample magnitude is unity and thus the maximum butterfly output is 2. Any output greater than unity would cause an undetected overflow, and prescaling the input by a factor of 0.5 is advisable. Scaling, however, is an undesirable step, since it reduces the dynamic range by a factor of two. Example 5.5 gives an alternative DIF radix-2 butterfly algorithm in which *dynamic scaling* is used. Here, the 32-bit accumulator is loaded with the input, left-shifted by 14. This means that the high half of the accumulator contains the input

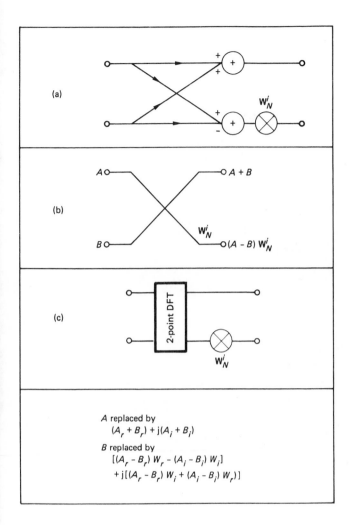

Fig 5.3 Three representation of the DIF butterfly:
   a) Flow graph representation
   b) Conventional butterfly symbol
   c) 2-point DFT model

EXAMPLE No. 5.3

Pascal Radix 2 DIF butterfly.

procedure butterfly2f (var x:signal; p,q:integer; wr,wi:real);

x is the signal array, p and q are the array indices. x[2*p] and x[2*q] are real parts, x[2*p+1] and x[2*q+1] are imaginary parts. wr and wi are real and imaginary parts of Wi.

var
        tr,ti   : real;   {temporary storage}

```
begin
      tr := x[2*p] - x[2*q];
      ti := x[2*p+1] - x[2*q+1];
      x[2*p] := x[2*p] + x[2*q];
      x[2*p+1] := x[2*p+1] + x[2*q+1];
      x[2*q] := tr * wr - ti * wi;
      x[2*q+1] := tr * wi + ti * wr;
end; {butterfly2f}
```

235

TMS EXAMPLE No. 5.4

TMS 320 DIF butterfly.

```
        LAC    XPR
        SUB    XQR
        SACL   TR              :TR = XPR - XQR

        ADD    XQR,1
        SACL   XPR             :XPR = XPR + XQR

        LAC    XQR
        SUB    XQI
        SACL   TI              :TI = XPI - XQI

        ADD    XQI,1
        SACL   XPI             :XPI = XPI + XQI

        LT     C               :C is WR
        MPY    TI
        LTP    S               :S is -WI
        MPY    TR
        SPAC
        SACH   XQI,1           :XQI = TI*C - TR*S

        MPY    TI
        LTP    C
        MPY    TR
        APAC
        SACH   XQR,1           :XQR = TR*C + TI*S
```

TMS EXAMPLE No. 5.5

TMS 320 DIF butterfly, dynamically scaled.

```
        LAC    XPR,14
        SUB    XQR,14
        SACH   TR,1            :TR = (XPR - XQR)/2

        ADD    XQR,15
        SACH   XPR,1           :XPR = (XPR + XQR)/2

        LAC    XQR,14
        SUB    XQI,14
        SACH   TI,1            :TI = (XPI - XQI)/2

        ADD    XQI,15
        SACH   XPI,1           :XPI = (XPI + XQI)/2

        .
        .
```

scaled by 0.25. No accuracy is lost since the full accumulator is used and no information is prematurely discarded. The results are stored from the high accumulator half with a left shift of 1, resulting in an overall scaling of 0.5. This process ensures that overflow will never occur, no matter how many butterflies are cascaded.

We have concentrated thus far on the DIF butterfly. For completeness, Fig 5.4 gives the pictorial representations of the radix-2 DIT butterfly. The output in this case is calculated from the input using

$$XP(out) = [XPR + (XQR \cdot WR - XQI \cdot WI)]$$
$$+ j[XPI + (XQI \cdot WR + XQR \cdot WI)]$$

$$XQ(out) = [XPR - (XQR \cdot WR - XQI \cdot WI)]$$
$$+ j[XPI - (XQI \cdot WR + XQR \cdot WI)]$$

Example 5.6 is part of the associated Pascal procedure. For DSP implementation, the choice of butterfly type, DIT or DIF,

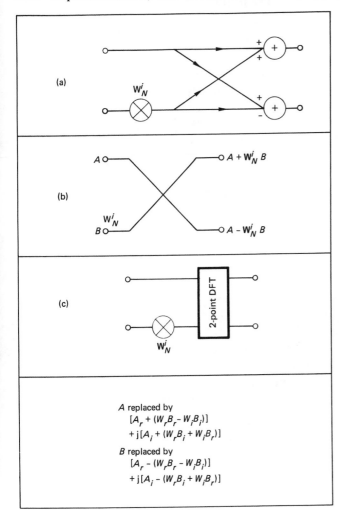

(a)

(b)

(c)

A replaced by
$[A_r + (W_r B_r - W_i B_i)]$
$+ j[A_i + (W_r B_i + W_i B_r)]$
B replaced by
$[A_r - (W_r B_r - W_i B_i)]$
$+ j[A_i - (W_r B_i + W_i B_r)]$

Fig 5.4 Three representations of the radix-2 DIT butterfly:
   a) Flow graph representation
   b) Conventional butterfly symbol
   c) 2-point DFT model

EXAMPLE No. 5.6

Pascal Radix 2 DIT butterfly.

```
begin
        tr := x[2*q] * wr - x[2*q+1] * wi;
        ti := x[2*q+1] * wr - x[2*q] * wi;
        x[2*p] := x[2*p] + tr;
        x[2*p+1] := x[2*p+1] + ti;
        x[2*q] := x[2*p] - 2*tr;
        x[2*q+1] := x[2*p+1] - 2*ti;
end; {butterfly2t}
```

depends largely on the instruction set of the device, and the overall application—one often proving more efficient than another.

5.3.3 The Cooley–Tukey radix-2 DFT

When the length of a DFT is a power of 2, it is possible to do the calculation solely with radix-2 butterflies. The method consists of decomposition of the N-point transform into two $N/2$-point transforms, as has been illustrated in section 5.3.1, but then continuing by decomposing each of the two $N/2$-point transforms into $N/4$-point transforms and so on, until only two-point transforms remain. Figs 5.5 and 5.6 illustrate the process for DIF and DIT approaches respectively. It can be shown that the FFT involves approximately $N \log_2 N$ complex multiplications compared with N^2 for direct implementation of the DFT. Since DSP devices can typically perform additions and multiplications in the same number of clock cycles, the number of additions involved in the FFT is also significant. A comparison of the relative execution times for the DFT and FFT can be found in section 5.4.

It is apparent from Figs 5.5 and 5.6 that the 'in-place' property is retained for any number of decomposition stages. The array used to store the input is now overwritten after each stage.

One further feature of the complete FFT algorithm is the rearrangement of the element indices. For a single decomposition into two half-length DFTs the scrambling consists of even index elements followed by the odd index elements. However, in the full decomposition for the FFT, repeated odd/even scrambling results in an overall reordering which corresponds to *bit reversal* of indices. This reordering is discussed further in section 5.3.4.

5.3.4 FFT input/output reordering

Table 5.2 illustrates the steps involved in reordering the output elements of the DIF algorithm. It is not always necessary to reorder the output, as for example when just a single component is needed, or when the DFT results (after some frequency domain processing) are to be input to an associated IDFT routine.

Note that the scrambling/unscrambling operation is just a sequence of operations in which the contents of pairs of memory locations are exchanged. In DSP device implementations when reordering is unavoidable, it is usually done by repeated execution of a block of code which performs a single exchange. Instead of using a loop for this purpose, it is preferable, if sufficient program memory is available, for the code itself to be repeated. This avoids the additional overheads associated with conditional branch instructions for loops.

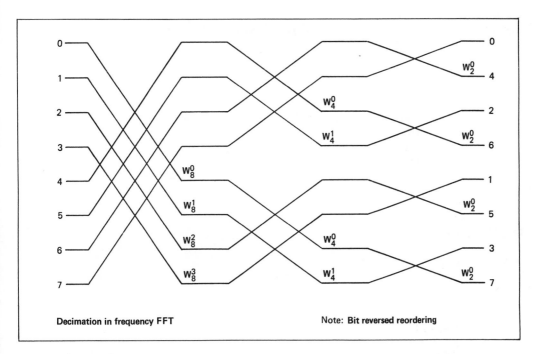

Decimation in frequency FFT

Note: **Bit reversed reordering**

Fig 5.5 Cooley-Tukey radix-2 FFT formed by decomposing the DFT into pairs of half-length DFTs using DIF decomposition

Fig 5.6 Cooley-Tukey radix-2 FFT formed by composing the DFT from pairs of half-length DFTs using DIT decomposition

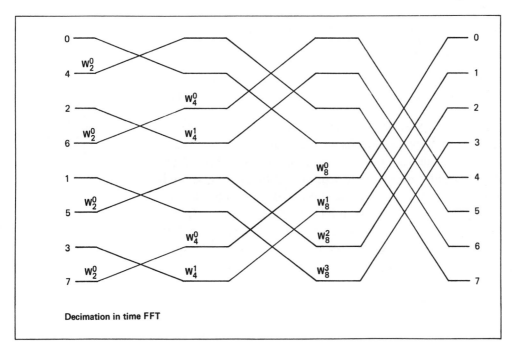

Decimation in time FFT

DIF	Array element position	Binary code	Bit reversed code	Corrected element position
	0	000	000	0
	1	001	100	4
	2	010	010	2
	3	011	110	6
	4	100	001	1
	5	101	101	5
	6	110	011	3
	7	111	111	7

Table 5.2 Bit reversal reordering

TMS EXAMPLE No. 5.7

TMS 32020 macro to swap storage locations.

Auxiliary register pointer should be set to AR1 on entry, and exit.

```
BITREV     $MACRO A,B,BIAS

           LRLK    AR1,:A:+:BIAS:
           LRLK    AR2,:B:+:BIAS:

           ZALH    *,AR2          :AR to high accumulator
           ADDS    *,AR1          :BR to low accumulator
           SACL    *+,0,AR2       :BR to former AR
           SACH    *+,0,AR1       :AR to former BR

           ZALH    *,AR2          :AI to high accumulator
           ADDS    *,AR1          :BI to low accumulator
           SACL    *-,0,AR2       :BI to former AI
           SACH    *-,0,AR1       :AI to former BI

           $END
```

Example 5.7 gives a TMS32020 macro definition for reversing the contents of a pair of memory locations whose addresses are passed as locations A and B on the data page, beginning with location BIAS. Again real and imaginary parts in consecutive locations are assumed.

The use of the macro for performing the reordering in an 8-point DIF is given in Example 5.8. It is assumed that the FFT has obtained and loaded its current data from locations 512 to 527 (0 to 15 on data page 4). The symbols X0 to X7 are associated with data memory locations 0, 2, ... 14 of the current data page.

5.3.5 Twiddle factor storage

Another feature of the repeated decomposition in the FFT is that the

TMS EXAMPLE No. 5.8

TMS 32020 reordering code.

Define data memory map.

X0	EQU	0	:real part of 0th. sample
X1	EQU	2	:real part of 1st. sample
X2	EQU	4	:etc.
X3	EQU	6	
X4	EQU	8	
X5	EQU	10	
X6	EQU	12	
X7	EQU	14	

.
.

Perform reordering.

```
BITREV X1,X4,512
BITREV X6,X3,512
```
.
.

TMS EXAMPLE No. 5.9

TMS 320 twiddle storage.

SINE	DATA	0	:0
	DATA	23169	:0.707 in Q15 format
	DATA	32767	:0.999 in Q15 format
	DATA	23169	:etc.
	DATA	0	
	DATA	-23169	
	DATA	-32767	
	DATA	-23169	

magnitudes of the angles of the twiddle factors are always less than π radians. The twiddle factor information can be stored in memory as the sine of the magnitude of the twiddle angle, as demonstrated in Example 5.9. (Note that the imaginary part of the twiddle factor is in fact the negative of this sine value, but it is stored in this fashion to allow the real part (cosine value) to be extracted by simply offsetting the sine index by $N/4$ positions.)

5.3.6 Programming the algorithm

Referring to Fig 5.5, the DIF FFT can be implemented with three nested loops. The outer loop covers the stages, the next loop covers each of the twiddle factors in a stage, and the innermost loop covers the butterflies associated with each twiddle factor. The algorithm is presented in structured English in Fig 5.7.

Table 5.3 shows how just two variables are needed to control the processing of each stage, each variable being halved on completion of the stage. Labelling the index of the upper input to the butterfly as P and the lower as Q, then the first butterfly involving twiddle Wi (Fig 5.5) has $P = i$, the next has $P = i + $ (twiddle period), etc., until all the input elements are exhausted. The term *wing span* is used for the variable $(Q - P)$.

Example 5.10 gives a complete Pascal procedure for the radix-2 DIF FFT, consistent with the variable names used above. The butterfly used is that demonstrated in Example 5.3 earlier. Table 5.4 gives the intermediate calculation results for the 8-point transform depicted in Fig 5.8. These results may be used to verify the correct operation of FFT software. Another useful test sequence is an all-zero input apart from a unity real part or imaginary part of one element. If this is the nth element, then the result $X(k)$ should be

---

Repeat (all stages)

    determine twiddle period
    determine butterfly span
    determine twiddles/stage
    determine butterflies/twiddle

    Repeat (all twiddles)

        start = 0
        calculate sin, cos

        Repeat (all butterflies)

            end = start + span
            butterfly(start,end,sin,cos)
            start = start + period

        Until all butterflies done
    Until all twiddles done
Until all stages done

Fig 5.7 The FFT algorithm in structured English

	Starting value	Next stage	Variable
Twiddle subscript	N	$\times 1/2$	N_1
Twiddle period	N	$\times 1/2$	N_1
Twiddles/ stage	$N/2$	$\times 1/2$	N_2
Wing span	$N/2$	$\times 1/2$	N_2

Table 5.3 Variation of key parameters in successive stages of FFT

$$\cos(2\pi kn/N) - \mathrm{j}\,\sin(2\pi kn/N)$$

for a real input element, or for an imaginary element,

$$-\sin(2\pi kn/N) + \mathrm{j}\,\cos(2\pi kn/N)$$

EXAMPLE No. 5.10

Pascal DIF FFT.

procedure fft2df(var x:signal; n:integer);

Radix 2 DIF FFT, length n must be a power of 2. Signal samples in x, real parts are in even numbered indices.

```
var
        p          : integer;          {butterfly input first index}
        q          : integer;          {butterfly input second index}
        wr,wi      : real;             {Re(W), Im(W)}
        n1         : integer;          {twiddle period of stage}
        n2         : integer;          {maximum twiddle index of stage}
        j          : integer;          {twiddle index}

begin
        n2 := n;
        repeat                         {all stages}
            n1 := n2;
            n2 := n2 div 2;
            j := 0;
            repeat                     {all twiddle indices}
                wr := cos (2*pi*j/n1);
                wi := -sin (2*pi*j/n1);
                p := j;
                repeat                 {all butterfly starts}
                    q := p + n2;
                    butterfly2f (x,p,q,wr,wi);
                    p := p + n1;
                until p >= n;
                j := j +1;
            until j = n2;
        until n2 = 1;
end; {fft2df}
```

Magnitudes after reordering:

Index		Magnitude
0		4.0
1	(from 4)	2.613 126
2	(from 2)	0.0
3	(from 6)	1.082 392
4	(from 1)	0.0
5	(from 5)	1.082 392
6	(from 3)	0.0
7	(from 7)	2.613 126

Table 5.4 Intermediate-stage outputs for 8-point DFT in Fig 5.8 (see also p. 244)

Input

Index	Real part	Imaginary part
0	1.0	0.0
1	1.0	0.0
2	1.0	0.0
3	1.0	0.0
4	0.0	0.0
5	0.0	0.0
6	0.0	0.0
7	0.0	0.0

After first stage 2-point DFTs:

Index	Real part	Imaginary part
0	1.0	0.0
1	1.0	0.0
2	1.0	0.0
3	1.0	0.0
4	1.0	0.0
5	0.707 107	− 0.707 107
6	0.0	− 1.0
7	− 0.707 107	− 0.707 107

After second-stage 2-point DFTs:

Index	Real part	Imaginary part
0	2.0	0.0
1	2.0	0.0
2	0.0	0.0
3	0.0	0.0
4	1.0	− 1.0
5	0.0	− 1.414 214
6	1.0	1.0
7	0.0	− 1.414 214

Output after final stage of DFTs:

Index	Real Part	Imaginary Part	Magnitude
0	4.0	0.0	4.0
1	0.0	0.0	0.0
2	0.0	0.0	0.0
3	0.0	0.0	0.0
4	1.0	− 2.414 214	2.613 126
5	1.0	0.414 214	1.082 392
6	1.0	− 0.414 214	1.082 392
7	1.0	2.414 214	2.613 126

Table 5.4 (continued)

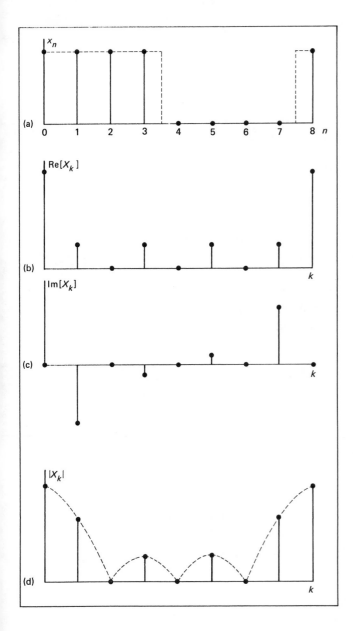

Fig 5.8 8-point DFT example:
  a) Input sequence
  b) Real part of transform
  c) Imaginary part of transform
  d) Resultant transform magnitude

5.3.7 Variations of the FFT algorithm

The radix-2 is by far the most commonly used, despite the fact that it suffers from the significant limitation of only working for sequence lengths that are a power of 2. Fast algorithms can in fact be developed for any value of N that can be expressed as a product of prime factors; however, the algorithms are specific to that particular value of N.

The principle upon which all the fast algorithms are based is the decomposition of an N-point transform into $N1$ transforms of length $N2$, followed by twiddling, followed by $N2$ transforms of length $N1$. The elements associated with the various transforms are found by arranging the indices in $N2$ columns of length $N1$. The first stage involves an 'in-place' DFT of each row, then twiddle, then an 'in-place' DFT of each column. If a is used as the column index, running from 0 to $N2$, and b is the row index running from 0 to $N1$, then the twiddle factor for element (a, b) is

$$W_N^{ab}$$

The results are scrambled such that the output indices associated with each position in the array must be found by writing them in sequence into the rows, then reading down the columns. For example, in Fig 5.9a, $N1 = 3$ and $N2 = 5$. The first stage consists of 5 three-point DFTs commencing with elements 8, 5 and 10. After twiddling there are 3 five-point DFTs commencing with 0, 1, 2, 3, and 4. The results are read out of the array in accordance with Fig 5.9b.

(a)	0	5	10	(b)	0	1	2
	1	6	11		3	4	5
	2	7	12		6	7	8
	3	8	13		9	10	11
	4	9	14		12	13	14

Fig 5.9 Prime factor 15-point DFT index mapping:
a) Input
b) Output

Another version of the FFT is due to Winograd. Compared with the Cooley–Tukey FFT, more additions are needed but fewer multiplications. Since, in general, DSP devices can multiply as fast as they can add, the Winograd FFT offers no advantage and will not be covered further.

5.4 FFT implementation using DSP

Having discussed the constituent parts of the FFT algorithm, we now consider methods for realizing the complete algorithm using DSP techniques.

5.4.1 An example program

Example 5.11 is a complete 8-point DIF FFT in TMS32020 code. The input data is assumed to start at data memory location 530. As far as possible the program structure matches the Pascal program shown in

Example 5.10, which itself is modelled on that in Fig 5.5 differing only in the use of a look-up table for twiddle factors, and down-counting loops.

The variables C and S are used for the real and negative imaginary parts of the twiddle factor, and K and L are used as downcounters for the stage and twiddle loops. The appropriate twiddle factors are extracted by means of a pointer (STP) which is incremented by STPI in the twiddle loop. The variable QMP is used for the wing span. This value is loaded into auxiliary register AR0 in order to provide an automatic offset from the P address, since the TMS32020 instruction set allows automatic incrementation of an auxiliary register by the contents of AR0.

The names of the remaining variables in the TMS320 program have been chosen to match those in the Pascal verion. Note, however, that P in the TMS320 program is the input data array pointer rather than the input complex element pointer. Data memory from 512 to 529 is used for program variables, and beyond this for the data array.

5.4.2 System memory and I/O considerations

In a typical real time application, while one block of data is being transformed, signal samples from the next block are being collected, and the results of the previous block are being output. One way of achieving this flow is to allocate two areas of memory, each large enough to accommodate N complex data values. While one is being used for the FFT, the other can be overwritten with new data as each individual processed sample has been output. A software flag can be used to indicate the current status of each buffer.

When the time taken to execute the FFT is much less than the time required to collect the data, more economical use of program memory is obtained if, instead of the double buffer, a circular buffer is used. Here, a pair of pointers takes care of data management, one marking the beginning of the current block being processed and the second marking the location of the next input sample. The data and pointers 'wrap around' on reaching the end of the buffer. The circular buffer must be large enough to accommodate one full block of data and as many samples as accumulate during the time taken to perform the FFT and output the results.

An important consideration in DSP devices which multiplex the external data and program buses is the need to accommodate as much of the data (and twiddle factors) in on-chip memory as is possible, since external memory fetches can, depending on the device used, require two cycles. At one extreme, when on-chip memory is very limited it may only be possible to store the data for one butterfly

TMS EXAMPLE No. 5.11

An 8 point FFT, implemented on the TMS 32020.

The example implements a radix 2 Cooley Tukey DIF FFT, using a single butterfly, with no reordering of data. Dynamic scaling is used. The input data is in complex form, stored in data ram with the real and imaginary parts in consecutive locations. Data is read into internal ram through input port PA1. The length limit is set by the memory size. A lookup table is used for twiddle factors.

	AORG	0	:set address for program code
	B	INIT	:send resets to initialisation
	B	INTRPT	:send interrupts to handler

Begin main program.

Declare symbols used to reference constants and data memory locations.

NVAL	EQU	8	:order of FFT
VALUES	EQU	15	:No. of data elements (real,im.)
M	EQU	3	
PG4	EQU	>200	:start address of data page 4
ONE	EQU	0	:holds 1
STPI	EQU	1	:sine table pointer increment
N	EQU	2	:length
N2	EQU	2	:max. twiddle superscript of stage
W4	EQU	4	:pointer offset cosine from sine
NCON	EQU	5	:number of values to initialise
P	EQU	5	:2 X butterfly input first index
Q	EQU	6	:2 X butterfly input second index
QMP	EQU	7	:2 X wingspan
C	EQU	8	:real part of twiddle factor
S	EQU	90	:imaginary part of twiddle factor
N1	EQU	10	:twiddle period of stage
I	EQU	11	:current twiddle superscript
STP	EQU	12	:sine table pointer
TR	EQU	13	:temporary store in butterfly comp
TI	EQU	14	:temporary store in butterfly comp
SI	EQU	15	:address of current element
DATST	EQU	32	:address of start of data

Place lookup table of sine values in program memory, for twiddle factors. Values in Q15 format. Increment = (2*pi)/N.

SINE	DATA	0,23169,32767,23169

```
        DATA    0,-23169,-32767,-23169
        DATA    0,23169
```

Place lookup table in program memory for initialising data memory.

```
TABLE   DATA    1                       :for ONE
        DATA    1                       :for STPI
        DATA    NVAL                    :for N
        DATA    NVAL                    :N2 = N
        DATA    N/4                     :W4 = N/4
```

Interrupt handler - interrupts not being used, so just a return.

```
INTRPT  RET                             :return
```

Initialisation routine.

```
INIT    DINT                            :disable interrupts
        SOVM                            :overflow on
        SSXM                            :sign extension on
        CNFD                            :block B0 is data memory
        SPM     0                       :no shift on P output
        LDPK    4                       :use data page 4
        LARP    AR1                     :point to AR1

        LRLK    AR1,ONE+PG4             :point to first destination
        RPTK    NCON-1                  :load repeat count
        BLKP    TABLE,*+                :transfer from program to data
```

Enter the main loop.

Input the data values

```
        LRLK    AR1,DATST+PG4           :point to data start
        RPTK    VALUES-1                :repeat for 16 values
        IN      *+,PA1                  :input the data
```

Use AR1 for number of stages loop counter (K).

```
        LARK    AR1,M-1                 :K = M - 1

KLOOP   LARP    AR2                     :point to AR2
        LAC     N2,15
        SACH    N1,1                    :N1 = N2
        SACH    N2                      :N2 = N2/2

        ZAC                             :clear accumulator
        SACL    STP                     :STP = 0
        SACL    I                       :I = 0
```

Use AR2 for number of twiddles counter (L).

```
        LAR     AR2,N2
        MAR     *-                      :L = N2 -1
```

```
LLOOP     LACK   SINE               :point to sine table
          ADD    STP                :add pointer
          TBLR   S                  :read S
          ADD    W4                 :add sine to cosine offset
          TBLR   C                  :read C

          LAC    STP
          ADD    STPI
          SACL   STP                :update STP by STPI

          LAC    I
          SACL   P,1                :P = 2 * I
```

Inner loop: accumulator holds test value P - 2*N.

```
ILOOP     LALK   DATST+PG4          :point to data start
          ADD    P                  :add P
          SACL   SI                 :SI points to data

          LAC    N2,1
          SACL   QMP                :QMP is wingspan
          LAR    AR0,QMP            :set up AR0

          LAR    AR3,SI             :AR3 points to data
          LARP   AR3                :point to AR3
```

Radix 2 DIF butterfly (auto scaling).

```
          LAC    *0+,14
          SUB    *,14
          SACH   TR,1               :TR = (X[P] - X[Q]) / 2

          ADD    *0-,15
          SACH   *+,1               :X[P] = (X[P] + X[Q]) / 2

          LAC    *0+,14
          SUB    *,14
          SACH   TI,1               :TI = (X[P+1] - X[Q+1]) / 2

          ADD    *0-,15
          SACH   *0+,1              :X[P+1] = (X[P+1] + X[Q+1]) / 2

          LT     C
          MPY    TI
          LTP    S
          MPY    TR
          SPAC
          SACH   *-,1               :X[Q+1] = TI*C - TR*S

          MPY    TI
          LTP    C
          MPY    TR
```

```
        APAC
        SACH   *,1                     :X[Q] = TI*S + TR*C
```

End of butterfly. Output results and increment loop.

```
        LARP   AR2                     :point to AR2
        LAC    P
        ADD    N1,1
        SACL   P                       :P = P + (2 * N)
        SUB    N,1                     :accumulator = P - (2 * N)
        BLZ    ILOOP                   :repeat until P - (2 * N) >= 0
```

End of inner loop. Update L loop.

```
        LAC    I
        ADD    ONE
        SACL   I                       :I = I + 1
        BANZ   LLOOP                   :L = L - 1  until  L = 0
```

End of L loop. Update K loop.

```
        LAC    STPI,1
        SACL   STPI                    :STPI = STPI * 2

        LARP   AR1
        BANZ   KLOOP                   :K = K - 1  until  K = 0
```

Scrambled FFT completed.

```
STOP        B        STOP

            END                       :end of program
```

on-chip. Prior to each butterfly this data is loaded from external memory along with the appropriate twiddle factor from program memory. Following the butterfly, the data just processed is output and the next set fetched.

When on-chip memory is larger, but still not large enough to accommodate the full data block, it is advantageous to decompose the transform into blocks compatible with the size of on-chip memory or smaller. Suppose for example that a 512-point transform is to be executed on a TMS32020, configured so that only 512 words of on-chip memory are available. The transform is too large to be executed directly (remember there are 512 complex data points), but it can be done using the three-step DIT process implicit in Fig 5.1. (The DIF process in Fig 5.2 is of course equally suitable.) A tutorial example for developing the 512-point FFT is given in Appendix 2.

5.4.3 Linear code

Because branch instructions consume extra machine cycles, it is usually better to repeat code than use a loop or subroutine call (program memory space permitting). A linear coded version of the 8-point DIF FFT is given in Example 5.12. It is assumed that a macro BUTFLY has been defined which accepts the P and Q index addresses and also the address of the cosine of the twiddle factor. (Cosine and sine values are in consecutive locations.) The sequence of butterflies follows Fig 5.2.

Even faster execution could be achieved with a linear code program if the twiddle factors were directly incorporated in the code through use of 'multiply immediate' instruction. In the TMS32020 the penalty of using this instruction is reduced coefficient accuracy (13 bits instead of 16).

TMS EXAMPLE No. 5.12

TMS 320 linear code 8 point DFT.

Define data memory map.

XO	EQU	0
X1	EQU	2
X2	EQU	4
X3	EQU	6
X4	EQU	8
X5	EQU	10
X6	EQU	12
X7	EQU	14
C0	EQU	16
C1	EQU	18
C2	EQU	20
C3	EQU	22

Macro butterfly definition placed here.

Perform the operation.

```
BUTFLY X0,X4,C0
BUTFLY X1,X5,C1
BUTFLY X2,X6,C2
BUTFLY X3,X7,C3
BUTFLY X0,X2,C0
BUTFLY X4,X6,C0
BUTFLY X1,X3,C2
BUTFLY X5,X7,C2
BUTFLY X0,X1,C0
BUTFLY X2,X3,C0
BUTFLY X4,X5,C0
```

5.4.4 Special butterflies

When linear code is used there are further opportunities for maximising the execution time. The butterflies associated with twiddle factors which have certain symmetries may be separately coded to fully exploit any possible simplification. The most obvious example is the zeroth twiddle factor, which is simply a multiplication by unity. The fact that all code associated with twiddle multiplication can be omitted means that a purpose-programmed zero twiddle butterfly will run much faster than a general butterfly. The same applies, although to a lesser extent, to twiddle angle magnitudes of $\pi/2$, $\pi/4$ and $3\pi/4$. The program in Example 5.12 could be improved if macros for special butterflies corresponding to these twiddle angles were defined as say BFY0, BFY45, BFY90 and BFY135. The code would then appear as in Example 5.13.

TMS EXAMPLE No. 5.13

TMS 320 improved DFT coding using optimal butterfly routines.

Perform the operation.

```
BFY0 X0,X4
BFY45 X1,X5
BFY90 X2,X6
BFY135 X3,X7
BFY0 X0,X2
BFY0 X4,X6
BFY90 X1,X3
BFY90 X5,X7
BFY0 X0,X1
BFY0 X2,X3
BFY0 X4,X5
```

5.4.5 Higher radices

Up to now we have concentrated on the radix-2 butterfly, which arose naturally out of the process of decomposing an N-point DFT down to two-point DFTs, where N is a power of 2. The two-point DFT was a desirable end point since it involves no multiplications. (In the associated butterfly the only multiplications are those associated with twiddling.)

This zero multiplication property of the 2-point DFT is shared with the 4-point DFT. When N is a power of 4, a speed advantage can be obtained by using a so-called radix-4 algorithm. The decomposition is carried out in the same way as outlined in section 5.3.7. As in the radix-2 case, it is possible to define a radix-4 butterfly which combines a 4-point DFT with multiplication by a quartet of twiddle factors. Fig 5.10 gives the pictorial representation of a radix-4 butterfly, and Example 5.14 a Pascal implementation. For a 16-point DFT a total of eight 4-point DFTs are needed, compared with thirty-two 2-point DFTs. Higher-radix butterflies and mixed-radix FFTs are also possible, but the restriction in allowable values for N, and the comparatively small further increase in speed, make the use of radices beyond 4 uncommon.

EXAMPLE No. 5.14

Pascal Radix 4 DIF FFT.

procedure fft4dif(var x:signal; n:integer);

Radix 4 DIF FFT, length n must be a power of 4. Signal samples in x, real parts are in even numbered indices.

```
var
        p          : integer;        {butterfly input first index}
        q          : integer;        {butterfly input second index}
        r          : integer;        {butterfly input third index}
        s          : integer;        {butterfly input fourth index}
        c1,s1      : real;           {Re(W1), -Im(W1)}
        c2,s2      : real;           {Re(W2), -Im(W2)}
        c3,s3      : real;           {Re(W3), -Im(W3)}
        n1         : integer;        {twiddle period of stage}
        n2         : integer;        {maximum twiddle index of stage}
        j          : integer;        {twiddle index}
        wi         : integer;        {index to sin in twiddle array}
        w0         : integer;        {inc in twiddle angle index}
        w4         : integer;        {index inc from sin to cos}
```

procedure butterfly4f(var x:signal; p,q,r,s:integer; c1,s1,c2,s2,c3,s3:real);

```
var
        tr1,tr2,tr3,tr4,ti1,ti2,ti3,ti4 : real;

begin
        tr1        := x[2*p] + x[2*r];
        tr2        := x[2*p] - x[2*r];
        tr3        := x[2*q] + x[2*s];
        tr1        := x[2*q] - x[2*s];
        ti1        := x[2*p+1] + x[2*r+1];
        ti2        := x[2*p+1] - x[2*r+1];
        ti3        := x[2*q+1] + x[2*s+1];
        ti1        := x[2*q+1] - x[2*s+1];
        x[2*p]     := tr1 + tr3;
        x[2*p+1]   := ti1 + ti3;
        x[2*q]     := (tr2 + ti4)*c1 - (ti2 - tr4)*s1;
        x[2*q+1]   := (ti2 - tr4)*c1 + (tr2 + ti4)*s1;
        x[2*r]     := (tr1 - tr3)*c2 - (ti1 - ti3)*s2;
        x[2*r+1]   := (ti1 - ti3)*c2 + (tr1 - tr3)*s2;
        x[2*s]     := (tr2 - ti4)*c3 - (ti2 + tr4)*s3;
        x[2*s+1]   := (ti2 + tr4)*c3 + (tr2 - ti4)*s3;
end; {butterfly4f}
```

```
begin
      n2 := n;
      w0 := 1;
      w4 := n div 4;
      repeat                           {all stages}
          n1 := n2;
          n2 := n2 div 4;
          j := 0;
          repeat                       {all twiddle indices}
              wi := j*w0;
              c1 := sc[wi+w4];
              s1 := -sc[wi];
              c2 := sc[wi*2 + w4];
              s2 := -sc[wi*2];
              c3 := sc[wi*3 + w4];
              s3 := -sc[wi*3];
              p := j;
              repeat                   {all butterfly starts}
                  q := p + n2;
                  r := q + n2;
                  s := r + n2;
                  butterfly4f (x,p,q,r,s,c1,s1,c2,s2,c3,s3);
                  p := p +n1;
              until p >= n;
              j := j + 1;
          until j = n2;
          w0 := w0*4;
      until n2 = 1;
end; {fft4dif}
```

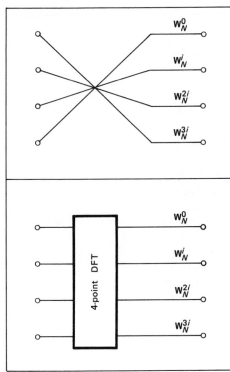

Fig 5.10 Radix-4 DIF butterfly representation

5.4.6 Exploiting special instructions and architectures

The examples given thus far have been based on the TMS320 family which are typical of general-purpose DSP ICs. Certain processors, however, have architectures or particular instruction sets designed specifically to increase the speed of FFT processing. Dual memory, for example, means that a single pointer can be used for both real and imaginary parts of signal samples. Where there are two address buses, a simultaneous fetch is possible. Automatic modulo-N counting reduces the overheads in the management of FFT loops and address pointers. Some processors have special instructions for generating a bit reversed index.

5.4.7 Purely real transforms

When baseband signals are being processed, the imaginary component of all input signal components is zero, and correspondingly the real and imaginary parts of the spectrum exhibit even and odd symmetry about zero frequency respectively (see for example the real transform in Fig 5.8). This means that a single N-point DFT can actually simultaneously process a pair of baseband signals.

When an N-point DFT is performed on a baseband signal, the upper half of the output contains no information not already present in the lower-half elements. Thus special butterflies could be used for the last stage of a DIF FFT which calculate only the upper-leg output. For even greater efficiency it is also possible to perform an N-point DFT on a baseband signal by means of an $N/2$-point DFT (see Ref. 5.1).

5.4.8 Performance comparisons

It is apparent from the foregoing that once the processing task

Algorithm	Size	Type	Device	Clock	Time
radix-2	64	looped	32010	5 MHz	2.87 ms
radix-4	64	looped	32010	5 MHz	1.92 ms
radix-4	64	linear	32010	5 MHz	0.60 ms
radix-2	128	looped	32020	5 MHz	4.375 ms
radix-2	256	looped	32020	5 MHz	8.483 ms
radix-2	256	linear	32020	5 MHz	4.519 ms
radix-4	256	linear	32020	5 MHz	3.1102 ms
radix-2	1024	looped	32010	5 MHz	69.4 ms
radix-2	1024	linear	32020	5 MHz	31.8198 ms

Table 5.5   FFT execution time examples

is defined and the DSP device has been decided upon, there is considerable scope for tailoring the algorithm to exploit the particular features of the data and the instruction set of the device. When high speed is needed there are three basic approaches: linear code, multiple butterflies, and radix-4 or radix-8 decompositions. Each technique has its own associated penalty. Table 5.5 gives some comparisons between algorithms and devices. Further examples of DFT programs using DSP can be found in Ref. 5.2.

5.5 Using the discrete Fourier transform to process continuous signals

We now have developed sufficient knowledge regarding the implementation of the DFT to consider its practical application in the spectral analysis of signals. We begin by considering a well-behaved class of signals which are periodic, with period corresponding exactly to some sub-multiple of the sampling rate. We then proceed to look at the more likely case of a periodic signal with non-specified period, and finally, in section 5.7, consider the analysis of random non-periodic waveforms such as speech.

As defined in section 5.2, the DFT is essentially a mapping of one set of N complex numbers to another. We can interpret the DFT as simply evaluating the z-transform of a discrete sequence at a set of points equally spaced along the z-plane unit circle. This section extends the interpretation to examine how the DFT is related to the Fourier transform of an ideally sampled continuous signal whose impulse weights generate the input discrete sequence. The DFT is a powerful tool for deducing information about the spectral properties of continuous and discrete signals.

A mapping of one set of complex numbers to another arises with the Fourier transform of a periodically sampled signal. If the period is T_p and the sampling interval T_s, the resulting spectrum will consist of a string of impulses (because the time domain was periodic) spaced at $f_p = 1/T_p$, and it will be periodic in the frequency domain (because of the time domain sampling) with a period of $f_s = 1/T_s$. Thus one period of the time domain has associated with it, $N = T_p/T_s$ complex numbers, and one period of the frequency domain has $f_s/f_p = T_p/T_s = N$ complex numbers, the two being related by a Fourier transformation.

Using the theory and results from Chapter 2, it is easy to derive the relationship between the Fourier transform and the DFT. Here we simply state the relationship in the form of a Rule.

The output of an N-point DFT may be interpreted as giving the Fourier series coefficients (scaled by a factor N) of the continuous bandlimited periodic signal whose samples over one period form the DFT input.

The rule is illustrated in Fig 5.11. To give a precise meaning to the rule, the term bandlimited must be further explained, and the definition of the Fourier series coefficients and the DFT itself clarified.

If the DFT input is assumed to have been generated by sampling a signal at rate f_s, then the signal must be bandlimited to within $f_s/2$. By imposing this restriction there is only one continuous signal that could have produced the sequence of samples. The equivalent Fourier series coefficients are those of the complex exponential series associated with that signal. As explained in Chapter 2, these complex numbers are also the weights of the impulses in the Fourier transform of the signal. The magnitude of the positive kth harmonic coefficient gives half the amplitude of the sinusoid at that frequency and its angle gives the phase.

Confusion sometimes arises because in fixed-point DSP device implementations it is normal to use scaling in the DFT algorithm so that the output elements are $\leqslant 1$ when the inputs are $\leqslant 1$. The appropriate scaling factor is $1/N$, which results in the FFT output values corresponding to exponential Fourier series coefficients directly. In this book, the standard definition of section 5.2, excluding the scaling factor, is used throughout, unless otherwise specified.

It is worth emphasizing that the DFT outputs relate to a periodic *continuous* signal, even though information about that signal has been supplied by means of samples.

As an example, consider the discrete sequence 1, 0, -1, 0 whose DFT is 0, 2, 0, 2. Interpreting the input sequence as samples of a continuous signal taken at intervals of T_s we can recover the original continuous signal, given that it was bandlimited to within $f_s/2$. Using techniques developed in Chapter 2 it can be shown that the only signal bandlimited to this extent which could have generated these samples is

$$x(t) = \cos(2\pi t/4T_s)$$

Taking the Fourier transform gives

$$X(f) = \tfrac{1}{2}\,\delta(f - f_0) + \tfrac{1}{2}\,\delta(f + f_0) \qquad f_0 = 1/4T_s$$

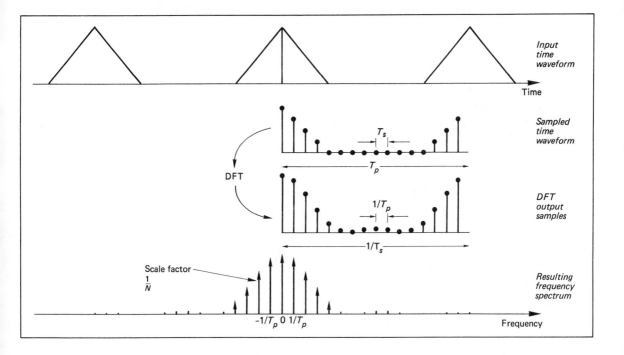

Labels in figure:
- Input time waveform
- Time
- Sampled time waveform
- T_s
- T_p
- DFT
- $1/T_p$
- DFT output samples
- $1/T_s$
- Scale factor $\frac{1}{N}$
- Resulting frequency spectrum
- $-1/T_p$ 0 $1/T_p$
- Frequency

which is equivalent to complex exponential Fourier series coefficients of magnitude 1/2 at frequencies $f_s/4$ and $-f_s/4$, and zero elsewhere.

How does this result relate to the output sequence from the DFT? To see the agreement, it is necessary to interpret the DFT index correctly. Scaling the DFT output by $1/N = 1/4$ as required by the rule gives

$$X(0)/4 = 0$$
$$X(1)/4 = 1/2$$
$$X(2)/4 = 0$$
$$X(3)/4 = 1/2$$

The sample number (index) gives the harmonic number of the associated positive Fourier series components directly for indices less than $N/2$. For the remainder, the harmonic number corresponding to negative Fourier series components is given by the index less $N/2$. The Fourier series coefficients yielded by the DFT are thus:

$$X_{-1} = 1/2 \qquad X_0 = 0 \qquad X_1 = 1/2$$

as expected.

To obtain the values for the trigonometric one-sided Fourier series components, we take only the first $N/2$ indices, and scale the sample values by two, e.g.

$$X_0' = 0 \qquad X_1' = 1$$

Fig 5.11 An interpretation of the DFT. If the numbers input to the DFT are samples of a periodic continuous signal then the output numbers represent the Fourier series coefficients of the original continuous signal multiplied by a factor N (provided the signal is sufficiently bandlimited)

5.5.1 Practical considerations when using the DFT

We have seen that, when the DFT is used to find the Fourier series coefficients of a bandlimited continuous periodic signal, the result is precisely correct (apart from an N times scaling factor) provided the DFT block length is an integer multiple of the signal period and N is chosen so that $N/2$ is higher than the highest non-zero harmonic. It is also precisely correct when used to find the Fourier series coefficients of a *sampled* periodic signal which meets the same conditions. However, the DFT result must be scaled by $1/NT_s = 1/T_p$ instead of $1/N$ and the corresponding spectrum is periodic with period $1/T_s$ (Fig 5.12).

The use of the DFT can be extended to other continuous signal processing tasks as indicated below, provided that the results are interpreted correctly. In some cases an error is introduced, but this can be made as small as desired by choosing a sufficiently large N.

Fig 5.12 An example of using the DFT to find the Fourier transform of an ideally sampled periodic signal

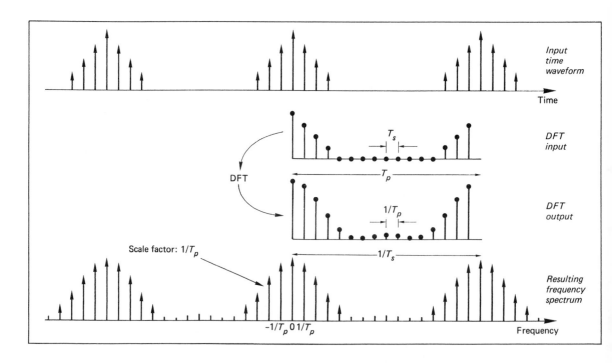

Case 1: **Non-periodic time-limited signals**
When a signal is of limited duration, e.g. a transient response, then the DFT can be used, provided the following condition is met:

The periodic version of the signal, formed by repeating the waveform indefinitely, must satisfy the bandwidth constraint

$$B \leqslant N/2T_p$$

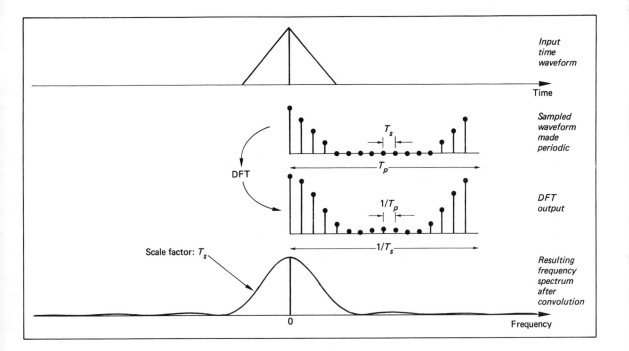

Fig 5.13 An example of using the DFT to find the Fourier transform of a bandlimited finite-duration continuous waveform. The output spectrum is found by convolving impulses spaced at $1/T_p$ (weights given by the DFT values), with the function sinc(fT_p) and a scaling factor of T_s

where T_p is the signal duration and N the length of the DFT. When the above condition is met, the spectrum of the original signal can be found by convolving the double-sided discrete frequency spectrum formed from the DFT output, with the sinc function

$$T_p \, \text{sinc}(fT_p)$$

Fig 5.13 illustrates the process involved. The accuracy of the resultant transform will depend on how valid the bandlimited assumption was, i.e. how much aliasing there is. Note that the correct scaling factor in this case incorporates a factor $1/N$ for converting from the DFT output to Fourier series component values, and a factor T_p introduced by the sinc convolution, giving a total scaling factor of $T_p/N = T_s$.

When the bandlimited condition does not apply, the spectrum will be affected by aliasing. Equivalently the spectrum yielded will be that of a slice of the unique periodic signal which is bandlimited and has the given samples.

If the original non-periodic signal was already sampled, it is then necessary to generate a periodic spectrum from the DFT output. This in turn is convolved with the sinc function as before, but there is an additional scaling factor of $1/T_s$ so the overall scaling is now unity (Fig 5.14).

Table 5.6 gives a complete summary of how the DFT can be used to analyse various categories of signal. In addition, it gives the

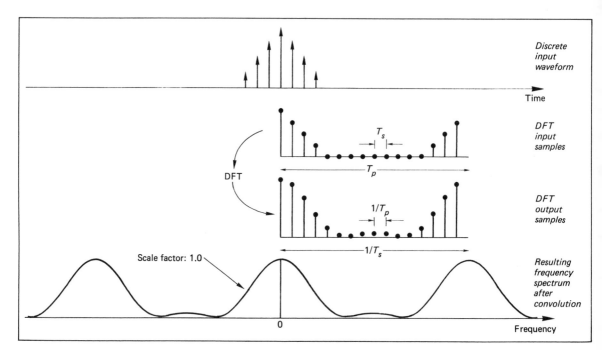

Fig 5.14 An example of using the DFT to find the Fourier transform of an ideally sampled finite-duration waveform (assumed to be bandlimited)

Fig 5.15 An example of using the IDFT to deduce the time domain response associated with a specified periodic ideally sampled spectral response

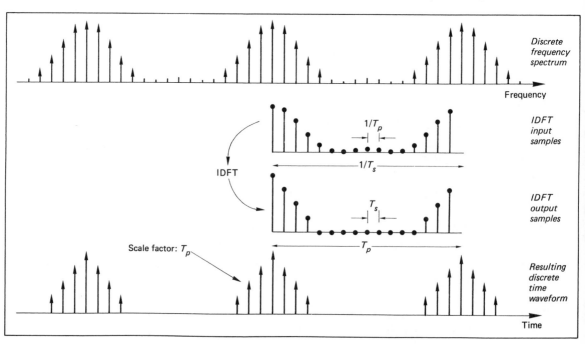

Waveform class	Continuous non-periodic	Sampled non-periodic	Continuous periodic	Sampled periodic
Scaling factor DFT	T_s	1	$1/N$	$1/T_p$
Scaling factor IDFT	$1/T_s$	N	1	T_p
Post-process	Convolve 1 period with sinc (fT_p) or sinc (tf_s)	Convolve with sinc (fT_p) or sinc (tf_s)	Select 1 period	None required

Table 5.6 Scaling factors for FT evaluation by summation

corresponding results for using the IDFT to infer the time domain properties of signals from various different categories of spectrum. Fig 5.15 is an example of the use of the IDFT to determine the time domain response, given the frequency response.

Case 2: **Periodic signals with non-related periods**
When the width of the signal slice used for the DFT does not correspond to the signal period (the practical case), then the signal whose Fourier series coefficients are found by the DFT will be a periodic repeat of the slice itself and thus may bear little resemblance to the original signal (Fig 5.16).

Fig 5.16 The waveform whose Fourier series coefficients are obtained when the DFT is applied to a non-integer multiple of a sinusoid period

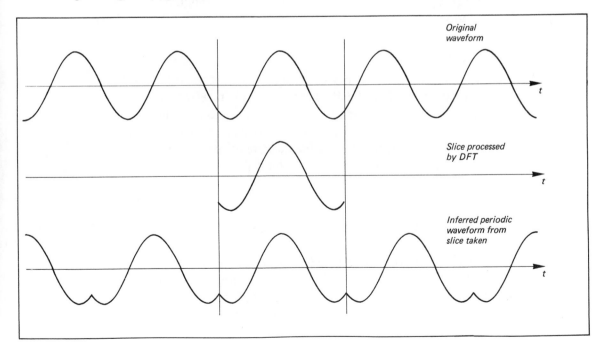

Original waveform

Slice processed by DFT

Inferred periodic waveform from slice taken

In order to interpret the resulting spectrum in this case, some of the techniques introduced in Chapter 2 must be used. The relationship between the DFT spectrum produced and the true spectrum is found as follows:

a) Convolve the DFT output samples with $W\,\text{sinc}(fW)$ where W is the slice width. (This gives the leakage introduced by windowing.)

b) Resample the resulting spectrum at intervals of $1/W$. (This takes account of the time domain periodicity implicit in the DFT operation.)

It is evident that making the slice encompass many periods of the original waveform minimises the scale of the error. The practical constraint here is the size of data buffer available, and the length of DFT that can be accommodated. There is also the possibility of controlling how the leakage error is distributed by choosing a suitable window shape. A phenomena that results from windowing prior to the DFT is known as the 'picket-fence' effect, often observed in DSP-based spectrum analysers. For a constant-magnitude input sinusoid with linearly increasing frequency the amplitude of the component in the spectrum analyser output is observed to increase, pass through a maximum, and then decrease before passing on to the next discrete frequency point available. The period of the fluctuation along the frequency axis is set by the size N of the DFT, and the magnitude controlled by the shape of the window. Fig 5.17 illustrates the picket-fence effect for two values of N, with both rectangular and triangular windows.

Case 3: **Stochastic signals**

Most real life signals fit into the category of stochastic processes, that is the signal cannot be predicted ahead of time, except in terms of statistical parameters. Section 5.7 of this chapter is devoted to the frequency domain analysis of this type of signal. However, before dealing with this topic, it is helpful to examine the role of the DFT in performing discrete time convolution and correlation.

5.6 Discrete time convolution and correlation

The FFT has greatly improved the efficiency of frequency domain processing of signals. In this section we investigate whether there is any advantage in deliberately 'migrating' into the frequency domain to perform operations which are naturally associated with the time domain. Two interesting examples are correlation and convolution.

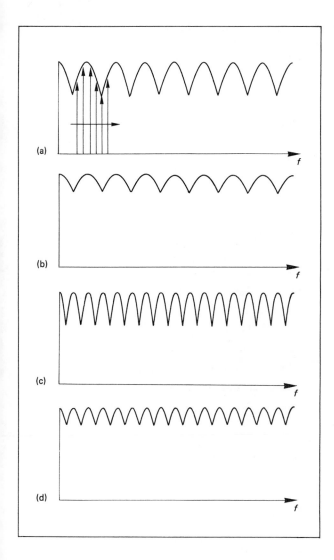

Fig 5.17 Illustration of the 'picket fence' effect, showing the DFT output assuming a constant-amplitude sinusoid input with frequency varying between zero and $f_s/2$:

    a) Rectangular window of width 16 samples

    b) Triangular window of width 16 samples

    c) Rectangular window of width 32 samples

    d) Triangular window of width 32 samples

We begin by defining the basic types of convolution and correlation for discrete signals.

5.6.1 Linear convolution of discrete signals

In Chapter 2, convolution was defined for continuous and discrete time signals and the equivalence of convolution in one domain and multiplication in the other was established. For *discrete signals* the equivalent results are:

Definition of Linear Convolution

$$y = x_1 \star x_2$$

implies

$$y(n) = \sum_{i=-\infty}^{\infty} x_1(i) \cdot x_2(n-i) \equiv Y(z) = X_1(z) \cdot X_2(z)$$

5.6.2 Aperiodic correlation of discrete signals

A closely related operation to convolution is correlation. The correlation of a signal x_1 with a second signal x_2 is defined as the time-reversed convolution of x_1 with a conjugated reverse running version of x_2. When x_1 and x_2 are the same we get the autocorrelation, otherwise the cross-correlation. The z transform of the autocorrelation is called the *power spectral density*, and of the cross-correlation, the *cross-spectrum*.

Definition of Aperiodic Correlation ϕ_{xy} and Cross-Spectrum G_{xy}

$$\phi_{x1x2}(n) = \sum_{i=-\infty}^{\infty} x_1(i+n) \star x_2{}^\star(i) = \sum_{i=-\infty}^{\infty} x_1(i) \cdot x_2{}^\star(i-n)$$

$$G_{x1x2}(z) = X_1(z) \cdot X_2{}^\star(z)$$

The processes of linear convolution and aperiodic correlation are illustrated in Fig 5.18.

5.6.3 Circular convolution of discrete signals

The definitions in the previous two sections are suitable only when at least one of the signals is restricted to a finite length, since otherwise the result is infinitely large. This unfortunately excludes a major class of signals, namely periodic signals. Since sinusoids of different frequencies are orthogonal, and all periodic signals can be expressed as the sum of harmonically related sinusoids, the only periodic signal pair giving a non-zero convolution result will be signals with the same period. For these particular class of signals, a second form of convolution can be defined which has a finite sum, called circular convolution.

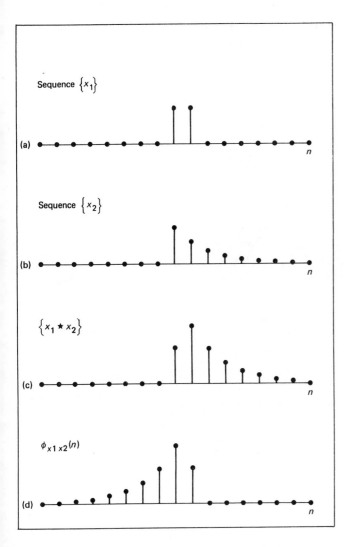

Sequence $\{x_1\}$

(a) ——————————————————— n

Sequence $\{x_2\}$

(b) ——————————————————— n

$\{x_1 \star x_2\}$

(c) ——————————————————— n

$\phi_{x1\,x2}(n)$

(d) ——————————————————— n

Fig 5.18 Illustration of linear convolution and aperiodic correlation:
  a) An arbitrary discrete signal
  b) An arbitrary discrete signal
  c) The linear convolution of *a)* and *b)*
  d) The aperiodic correlation of *a)* and *b)*

Definition of Discrete Circular Convolution
Given $x_1(n)$ and $x_2(n)$ each have length N, then

$$y(n) = \sum_{i=0}^{N-1} x_1(<i>) \cdot x_2(<n-i>)$$

where $<\ \ >$ indicates that the index is evaluated on a modulo N basis.

The result of the circular convolution $y(n)$ will be periodic with period N. The transform domain equivalent of this definition is

$$\text{DFT}\,[y(n)] = \text{DFT}\,[x_1(n)] \cdot \text{DFT}\,[x_2(n)]$$

or

$$Y(k) = X_1(k) \cdot X_2(k)$$

5.6.4 Periodic correlation and spectral density of discrete signals

The correlation of a pair of periodic discrete signals with the same period calls for a similarly modified definition. Periodic correlation is defined in terms of circular convolution in the same way that aperiodic correlation is defined in terms of linear convolution, except that it is conventional to use a scaling factor of $1/N$.

Definition of Periodic Correlation $R_{xy}(n)$

Given that $x_1(n)$ and $x_2(n)$ are each periodic with period N,

$$R_{x1x2}(n) = (1/N) \sum_{i=0}^{N-1} x_1(<i+n>) \cdot x_2(<i>)$$

where $<i>$ indicates $i \bmod N$ as before.

The function $R_{x1x2}(n)$ will also be periodic with period N. The transform domain equivalent is

$$\text{DFT}\,[R(n)] = \frac{1}{N} \cdot \text{DFT}\,[x_1(n)] \cdot \text{DFT}\,[x_2(n)]$$

5.6.5 Correlation of infinite length sequences

By allowing N in the periodic correlation definition to tend to infinity, a correlation definition is obtained that will give a finite value for an infinite length sequence. Non-periodic sequences of infinite length can only be described statistically. A measured estimate of the correlation function can be obtained from N samples using

$$\hat{R}_{x1x2}(n) = (1/N) \sum_{i=0}^{N-n} x_1(i+n) \cdot x_2\star(i)$$

where $x_2(i) = 0$ for $i < 0$.

The subject of correlation and spectral estimation of stochastic signals is taken up in section 5.7. We note in passing that the two definitions of discrete correlation just introduced mirror the two continuous signal correlation definitions given in Chapter 2, one for energy signals and one for power signals.

5.6.6 Convolution and correlation using the FFT

It would appear from the above discussion that the only operations in which the processing efficiency of the FFT could be exploited are circular convolution and periodic correlation, since only in these two cases does the transform domain equivalent involve the DFT. Fortunately there is a way around this apparent restriction. If $x_1(n)$ and $x_2(n)$ are aperiodic (without a natural period) with lengths L and M respectively, the length of the non-zero part of the convolution function will be $L + M$. The circular convolution definition will thus give the *same* result as linear convolution if the two input sequences are first extended with zeros to a length at least equal to the sum of their original lengths. Fig 5.19 illustrates how extending the period by zero padding makes the circular convolution between two periodic waveforms over one period, approach the same shape as the linear convolution of single periods of the two waveforms.

Fig 5.19 Illustration of the use of zero padding to make the circular convolution of a pair of periodic waveforms equivalent to the linear convolution over one period. The two input waveforms are periodic versions of those illustrated in Fig 5.18.

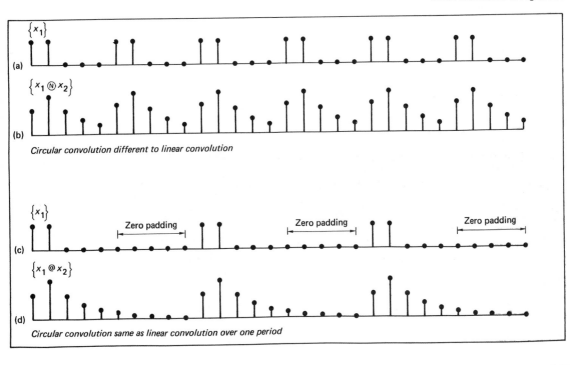

(a) $\{x_1\}$

(b) $\{x_1 \circledN x_2\}$

Circular convolution different to linear convolution

(c) $\{x_1\}$ Zero padding   Zero padding   Zero padding

(d) $\{x_1 \circledast x_2\}$

Circular convolution same as linear convolution over one period

The procedure for fast convolution is now clear:

a) Form an *N* length sequence from the longer of the two inputs, the first *N*/2 terms of which are the signal, with the last *N*/2 terms zero.

b) Form a second length *N* sequence from the second input, padding out with zeros to fill the *N* positions.

c) Find the DFT of each using the FFT.

d) Do a term-by-term multiplication of the DFTs.

e) Find the IDFT of the product. (This is the required convolution.)

The process is illustrated in Fig 5.20.

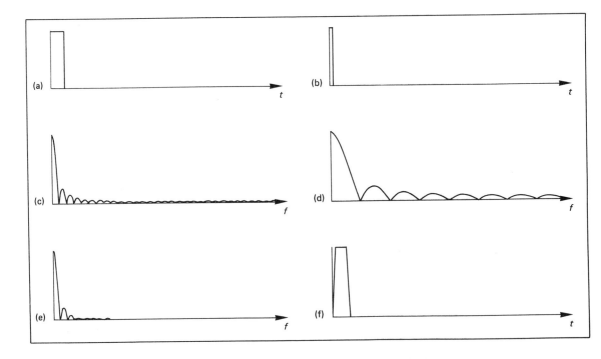

Fig 5.20 An example of convolution by Fourier transform:

a), b) Input waveforms to be convolved

c), d) Corresponding spectra after Fourier transform

e) The cross-spectrum [(c) × (d)]

f) The inverse Fourier transform of the cross-spectrum

This procedure will be precise, provided that the periodic signals, formed by repeating the length *N* padded sequences, are indeed strictly bandlimited to half the sampling rate. The zero padding does not involve any loss of information, it merely causes the DFT harmonics to be closer than strictly required by the sampling theorem. The zero padding does however incur a fourfold processing penalty between the DFT method and direct convolution. For large *N* this is more than offset by the speed advantage of using the FFT, which for this application requires no reordering, provided that an 'in-sequence' input FFT algorithm is used for the forward transform and a bit reversed sequence algorithm for the reverse transform.

The method outlined above applies equally well to aperiodic correlation. Here, a $1/N$ scaling factor is introduced, and the terms of the second signal DFT must be conjugated prior to step d.

In Appendix 3, tables are given summarising the methods available for deducing convolution, correlation and spectral density functions of various classes of input continuous signal from discrete operations on the signal samples, either directly or in the transform domain.

5.6.7 Filter implementation using the DFT

Since filtering is equivalent to linear convolution of the input sample with the filter impulse response, and we have just seen that fast linear convolution can be performed using the DFT, it might appear that a faster FIR filter implementation could be achieved using the FFT method than obtained by direct convolution. In fact this is only true for very large numbers of taps, the exact number depending on the way in which the FFT is implemented, and the processor being used.

The method is simply an extension of that used for performing linear convolution with finite length sequences, and is illustrated in Fig 5.21. The input signal is processed in slices of length $N/2$, where $N/2 \leqslant$ the length of the filter impulse response L. The convolution is performed with the aid of zero padding as before. The length of the non-zero section of the output will now be longer than $N/2$ by the factor L. In fact, the first L terms and the last L terms of the non-zero section will be transient terms which need to be added to the corresponding terms of the previous and following slice outputs to simulate convolution with a continuous signal. The output block of $N/2$ elements is thus formed from the $N/2 + L$ length sequence by taking the first L terms and adding these to the corresponding terms of the previous slice output.

A disadvantage of the DFT method of filtering is that a delay of at least the slice length is introduced and there are overheads involved in assembling the slices and outputting the results, compared to direct convolution where one output is produced each sampling period. For baseband FIR filters (real signals convolved with real impulse responses) the DFT method becomes attractive on the TMS32010 for filters with more than 100 taps.

5.7 Spectral analysis

In this section we consider DSP techniques for estimating the spectrum of random signals. Each appearance, or each slice, of a random signal is different; thus it is not satisfactory to simply find the Fourier transform of the sample at hand. If the signal is

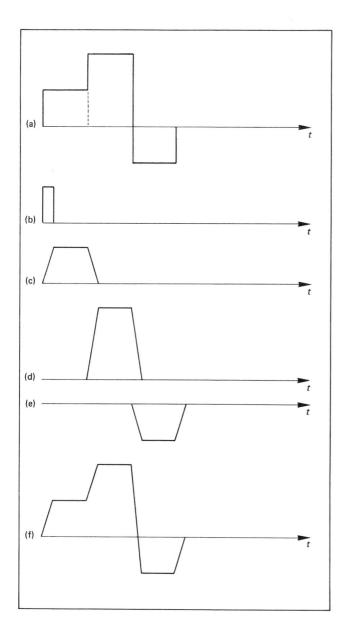

Fig 5.21 An example of filtering using block Fourier transforms:
  a) Input waveform
  b) Filter impulse response
  c), *d*), *e*) Components of output
  f) Resulting filtered output

stationary, it can be modelled as the output of a filter whose input is white noise. The power spectral density of the signal can then be determined as the square of the magnitude of the model filter transfer function.

The fundamental problem of real time spectral analysis is that only a finite length slice of signal is available, either because it remains stationary only for a limited time (like speech) or because of memory restrictions of the processor. The spectral information derived from the slice can therefore only be an estimate of the true power spectral density. Four methods for deriving the spectral density are considered here, two making use of autocorrelation function estimates, the third using frequency domain techniques exclusively, and a fourth approach whereby the pole locations of an *n*-pole filter are derived, which when operating on white noise would produce a signal most resembling that being analysed.

5.7.1 Block autocorrelation method

The first process in the block autocorrelation method is to take a slice of the input signal and multiply it with a time domain window function. A variety of different window shapes have been proposed, each with its own merits. The purpose of the windowing operation is to control leakage (see Chapter 2). A rectangular window, for example, minimizes the leakage close to a dominant spectral component, but results in much greater leakage at large frequency offsets compared with, say, a triangular window.

The next step in the spectral analysis is to perform a linear autocorrelation of the signal slice. This can be done either directly or by the DFT method described in the previous section. The greater the number of samples in the slice, the better the autocorrelation function estimate.

Before taking the DFT of the autocorrelation function (acf) estimate to obtain the power spectral density function (cf. section 2.6.4), it is necessary to perform an intermediate step of preferentially weighting (windowing) the acf values associated with small relative time shifts. The simplest form of acf weighting is to set to zero all acf estimates for delays with magnitude $\geqslant T_w$. This is equivalent to the frequency domain operation of convolving the original spectral estimate with a $\text{sinc}(fT_w)$ characteristic. Such a process occurs automatically in a swept frequency spectrum analyser where the actual spectrum displayed is the convolution of the IF filter characteristic with the true input spectrum. A narrow window on the acf is equivalent to a wide bandwidth filter in a spectrum analyser, thus giving maximum smoothing. If no window is used, the DFT in effect gives the power output of very very narrowband filters.

Because the output of a particular filter will vary considerably from slice to slice (even though on average it will give the correct power spectral density), it is necessary to pool the outputs of adjacent filters to get a smooth spectrum. There is a trade-off between smoothness on the one hand and resolution and lack of bias on the other, controlled by the width of the acf window.

In Fig 5.22 the spectral estimation of a slice of a random signal using this method is illustrated. The signal was generated by passing

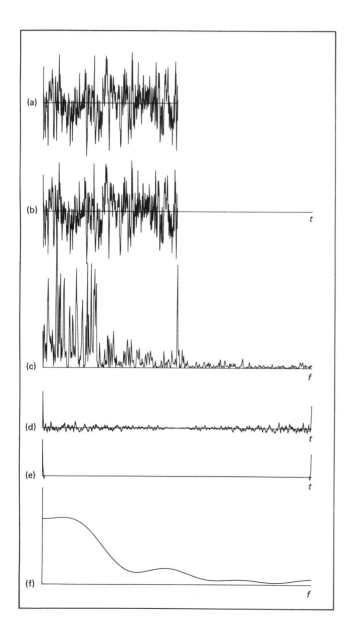

Fig 5.22 An illustration of the block autocorrelation method of spectral analysis:
    a)   Input time waveform
    b)   Input with zero padding
    c)   Cross-spectrum
    d)   Autocorrelation function (acf)
    e)   Windowed acf
    f)   Spectral estimate

white noise through a first-order filter with cut-off at $f_s/4$ and adding a low-level sinusoid at $f_s/2$. The steps in the processing are firstly zero padding, then finding the DFT and the cross-spectrum (Fig 5.22c), then the autocorrelation function (Fig 5.22d), then windowing the autocorrelation function (Fig 5.22e), and finally taking the DFT of the windowed autocorrelation function to give the power spectral density. In Fig 5.22 the acf window is only 16 samples wide, compared with the slice width, including zero padding, of 1024

274

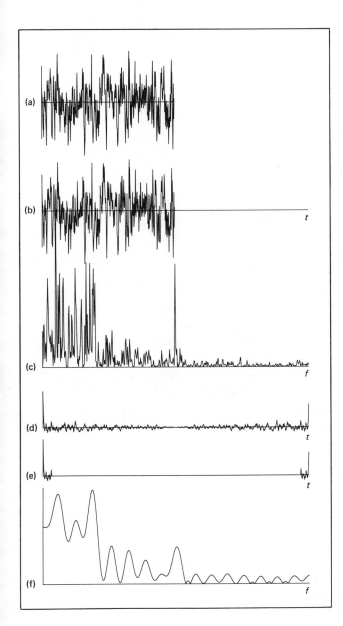

(a)

(b)   *t*

(c)   *f*

(d)   *t*

(e)   *t*

(f)   *f*

Fig 5.23 As for Fig 5.22 but with a window encompassing 64 out of 1024 samples compared with only 16 out of 1024 samples in Fig 5.22

samples. As a result a smoothed spectral estimate is obtained, so much so that the impulse associated with the embedded sinusoid is smeared out to occupy a wide bandwidth and is no longer distinguishable from the noise spectrum. In the cross-spectrum, on the other hand, the sinusoid is clearly visible, but the noise spectrum, although having the correct average level, shows large fluctuations from frequency to frequency.

In Fig 5.23 the same waveform is processed using a greater-width acf window (64 samples). The trade-off between smoothness and lack of bias (i.e. the extent to which the average power at a particular frequency will be correct if the spectrum of a large number of slices were to be separately estimated) is evident. Note how the sinusoid impulse is now more distinguishable.

5.7.2 Running autocorrelation function method

The acf is, in effect, an average of the product of samples spaced t seconds apart for all pairs of samples. When each new sample arrives it contributes one further set of products to the running averages associated with each delay. Thus, rather than taking a slice of the signal and processing the slice, the signal could be passed through a circular buffer, so that, as each new signal sample comes in, it can be multiplied by each appropriate stored delayed sample and the product used as the input of a very narrow bandwidth digital filter to provide the smoothing. One such filter will be needed for each delay of interest, but the filters can be recursive.

The advantage of this scheme over the block acf calculation is that the reliability for all delays is the same (they are all averaged over the same number of samples) and an acf estimate is available each sampling period. This means that the acf will gradually adapt to spectrum changes.

The number of autocorrelation function samples that need to be estimated will once again depend on the desired smoothness of the spectral estimate.

5.7.3 Spectrogram method

Another way of obtaining a smooth spectral estimate is to average the square of the magnitude of the DFT (or the product of the two DFTs after conjugation in the case of the cross-spectrum) over a large number of slices. For a given total block length there is once again a trade-off between resolution (small number of large slices) and smoothness (large number of small slices).

5.7.4 LPC coefficient extraction

In an FIR filter, the output is synthesized as a weighted sum of previous inputs. The filter taps are sometimes called *linear prediction coefficients* since they associate a future output contribution with the current state. The information contained in the LPC coefficients is the same as that in the zeros of the filter, and in the reflection coefficients of the lattice filter described in Chapter 4.

One way of describing the power spectrum of a signal with a specified level of precision is to find the filter with a predefined number of poles which will cause white noise to have an autocorrelation function which most closely matches the actual signal. A set of simultaneous equations can be set up in terms of the LPC coefficients, which, when solved, result in the appropriate filter specification for this task. Levinson (Ref. 5.3) first proposed a simple algorithm for solving these equations with a fast recursive algorithm. It has since been refined by others with a version of particular interest for fixed-point DSP devices developed by LeRoux and Gueguen (Ref. 5.4). This version gives the reflection coefficients (the K_i in the lattice realization of Fig 4.32) rather than the LPC coefficients, since these (and all intermediate results in the algorithm) are bounded within the range -1.0 to 1.0.

An example of the use of linear prediction processing is in speech synthesis. Here the LPC coefficients (or reflection coefficients) are stored and the sound synthesized by passage of white noise through the inverse of the whitening filter characterised by the LPC coefficients. This will be an all-pole filter which is best synthesized with a lattice structure. Algorithms for extracting the LPC coefficients and reflection coefficients from the signal slice autocorrelation function are given in Appendix 4.

5.7.5   Pitch estimation

This specific task of spectral analysis is important in the processing of speech and music signals. The objective is to find the period of a signal rich in harmonics, which may have many zero crossings per period. One way of estimating the pitch is to set a variable threshold at a level which results in the times between threshold crossing and subsequent zero crossing being approximately uniform, or at least having a uniform pattern of time intervals. These raw pitch interval estimates may then be processed over several samples to smooth the estimate.

Another approach is to use *homomorphic processing*. Because an arbitrary periodic signal can be modelled by a periodic impulse stream passing through a specified filter, the spectrum of a periodic signal can be expressed as

$$X(f) = \sum_{k=-\infty}^{\infty} \delta(f - kf_0) \cdot H(f)$$

$X(f)$ can be calculated using the DFT (it will only be an approximation since the DFT period will not generally coincide with the true period). Taking the logarithm converts the product to a sum, and the logarithm of the impulse train will still be periodic in the frequency domain with period f_0, whereas the log of $H(f)$ is aperiodic. Thus, taking the complex log of the DFT or even just the log of the magnitude, and then finding the IDFT of the result, will give time domain spikes spaced at $1/f_0$. The log of the DFT is called the *cepstrum* (Ref. 5.5).

References

5.1 Parks, T. W. and Burrus, C. S., *DFT/FFT and Convolution Algorithms*, Wiley (1985).

5.2 *Digital Signal Processing Applications*, Texas Instruments (1986).

5.3 Levinson, N., The Weiner rms error criterion in filter design and prediction, *J. Math. Phys.* 25 261, 278 (1947).

5.4 LeRoux, J. and Gueguen, G., A fixed-point computation of partial correlation coefficients, *IEEE Intern. Conf. on Acoustics, Speech and Signal Processing* (1977)

5.5 Rabiner, L. and Gold, B., *Theory and Application of Digital Signal Processors*, Prentice Hall (1975).

6 General signal processing algorithms

6.1 Introduction

Following on from the last two chapters which have dealt with specific algorithms and techniques for filtering and spectral analysis, this chapter provides a more general set of algorithms including techniques for waveform generation, modulation, demodulation and signal averaging. Reference is made to material presented in the previous three chapters as indicated in the text.

6.2 Waveform generation

Waveform generation is fundamental to communications, instrumentation, and control systems alike. Digital implementation of waveforms can provide a much greater accuracy, stability and flexibility than conventional analog implementations, permitting tighter specification and matching of systems and circuits.

When considering algorithms for waveform generation, of which there are several, the following points should be born in mind.

1 Waveform purity How pure does the waveform need to be? The purity of a sinewave is usually expressed in terms of Total Harmonic Distortion (THD), defined as

$$\text{THD} = \frac{\text{Power in unwanted harmonic components of waveform}}{\text{Total waveform power}}$$

For example, a sinewave generator with a THD of 0.1% has a distortion power level approximately 30 dB below the fundamental component.

2 Frequency/period control Does the frequency/period of the waveform need to be varied, e.g. voltage-controlled oscillator?

3 Phase control Does the relative phase of the waveform need to be varied?

4 Memory usage How much memory is required for waveform generation? (Often there is a trade-off between waveform purity, i.e. low THD, and memory requirement.)

5 Execution time What is the constraint on algorithm execution time? (Often there is a trade-off between execution time and memory requirement.)

6 Accuracy What is the tolerance required on the gain and frequency setting?

Each of the above factors are addressed for the specific waveform generation algorithms described in the following. The salient points are summarised in Table 6.1 to aid the choice of algorithm for a given application.

6.2.1 Look-up table/sampled waveform generator

This method of waveform generation is probably the most flexible and conceptually simple method of generating a given periodic waveform characteristic but is often overlooked by engineers with a background of analog design, as it is unique to digital signal processing and, in particular, mass data storage and recall facilities. The technique simply involves the reading out of a series of stored data values representing discrete samples of the waveform to be generated. The data values can be obtained either by sampling the appropriate analog waveform, or more commonly by computing the desired sample values.

Provided that enough samples are stored to accurately represent one complete period (less if the waveform is symmetrical) of the waveform, then continuous signals are generated by repeatedly cycling through the data memory locations. Variable frequency/period is readily achieved by either varying the rate at which the successive samples are read from memory (i.e. varying the program execution time), or by selecting one out of every N samples with a fixed program execution time. This latter process is subject to the constraints imposed by aliasing, requiring at least two samples per period of the highest frequency component in the desired waveform. The techniques involved in algorithm development are best understood by considering the important case of a sinewave generator.

Sinewave generator: look-up table method Fig 6.1 shows a sinewave represented by eight sample values equally spaced in time over one period of the waveform. The values are easily computed by evaluating the function

$$X(i) = \sin(i \times 360°/N)$$

Oscillator type	Harmonic content	Algorithm complexity	Program memory requirement	Phase/ frequency control	Execution time
Look-up table	LOW/ MEDIUM	LOW	HIGH	GOOD	LOW
Relaxation	HIGH	MEDIUM	LOW	AVERAGE	MEDIUM
Recursive	LOW	LOW	LOW	POOR	LOW

Table 6.1 Summary of algorithm characteristics for sinewave generation

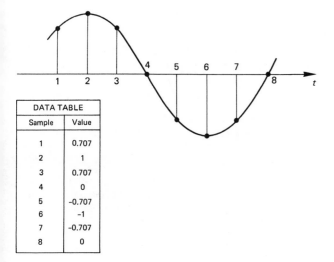

Fig 6.1 Sampled sine-wave (8 samples/period)

DATA TABLE	
Sample	Value
1	0.707
2	1
3	0.707
4	0
5	−0.707
6	−1
7	−0.707
8	0

where $X(i)$ is the ith sample value and N is the number of samples in a period (eight in this example). These sample values must now be represented in binary form, the accuracy being determined by the data word length, typically 16-bits. Once the data values are resident in memory, the sinewave is generated by cycling through the memory locations in turn and reading the stored value, the cycle rate and overall program execution time determining the frequency of oscillation. The address counter required to cycle through the memory locations can be any of those described in Chapter 3, a modulo counter usually proving the most efficient with wrap-around at the end of the data table automatically realized. Clearly, placing the waveform sample values in consecutive memory locations simplifies the addressing procedure significantly.

The drawback with the above technique as it stands is that the frequency of the example sinewave is fixed at exactly 1/8th of the program cycle frequency. To vary the oscillator frequency one has either to change the program execution rate or to miss out some of the stored samples. For example, incrementing the address counter by two rather than one means that every other waveform sample is read out from the data table, and the effective oscillator frequency is doubled with only four samples per period. (Note that at least two samples per period must be used to satisfy the Nyquist sampling constraint (Chapter 2) for a reproducible signal.)

By continuously varying the step size of the address counter, any frequency can be generated within the sampling constraints; however, the waveform accuracy can vary considerably. For example, if, using the eight stored data values, we require a signal with a frequency of 1/6th of the program execution rate, then the address counter must cycle through the complete waveform data table in 6 program cycles. Fig 6.2 demonstrates that, in this case, only two out of the eight stored samples will correspond to the correct value for the waveform, with the other four samples being approximated from the six remaining stored samples. This approximation is the cause of the waveform inaccuracy. Clearly, the more samples stored in data memory, the greater the choice and hence the less the discrepancy between the wanted and available sample values used by the waveform generator.

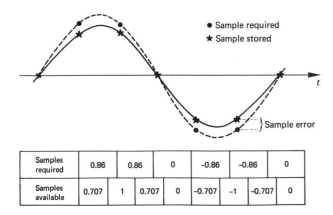

- Sample required
- Sample stored

} Sample error

Samples required	0.86	0.86	0	-0.86	-0.86	0
Samples available	0.707	1	0.707	0	-0.707	-1	-0.707	0

Fig 6.2 An example of waveform distortion due to a finite number of stored samples

The frequency of oscillation using the look-up table generator is given by the equation:

$$f = \text{STEP}/(T_s \times N)$$

where STEP is the address counter increment, T_s is the program cycle time, and N is the number of stored samples/cycle. The desired value of STEP in many cases will not be an integer (8/6 in the above example), and in order to derive the necessary integer address pointer value, truncation or rounding must be used.

The resolution to which the oscillator frequency can be set using this technique is clearly dependent on the accuracy to which the

STEP value can be represented using binary arithmetic. The maximum frequency error is in fact given by

$$\delta f_{max} = 1/(T_s \times 2^M)\ \text{Hz}$$

where M is the word length of the register containing STEP.

A sinusoidal waveform generator algorithm for the Texas Instruments TMS320-20 using this direct table look-up method is given in Example 6.1. This algorithm uses 32 stored data values and 16-bit arithmetic and has an execution time of 2.8 microseconds. (Note: the execution time is independent of the size of the data table.) The minimum frequency resolution is thus 5.4 Hz and the maximum oscillator frequency 178.6 kHz (keeping within the Nyquist constraint). The waveform accuracy for various integer and non-integer step sizes is shown in Table 6.2 as a function of the number of stored samples N, and serves to provide a quick guide to the minimum accuracy that can be expected for a given memory size.

Number of samples N	Step size	Total Harmonic Distortion (dB below fundamental)	
		Direct table look-up	Linear interpolation
	2	>60	>60
	2.25	25	28
	2.5	26	29
32	2.75	25	28
	3	>60	>60
	8.25	25	28
	11.625	25	28
	2	>60	>60
	2.25	31	37
	2.5	32	38
64	2.75	31	37
	3	>60	>60
	8.25	31	37
	11.625	31	37
	2	>60	>60
	2.25	37	>60
	2.5	38	>60
128	2.75	37	60
	3	>60	>60
	8.25	37	>60
	11.625	37	58

Table 6.2 Total harmonic distortion for look-up table sinewave generator

283

TMS EXAMPLE No. 6.1

Sinewave generation using a lookup table (length 32).

Place the lookup table in low program memory, in Q15 notation.

```
            AORG   >20                    :place after vectors

STABL       DATA   0,6393,12540,18205
            DATA   23170,27246,30274,32138
            DATA   32767,32138,30274,27246
            DATA   23170,18205,12540,6393
            DATA   0,-6393,-12540,-18205
            DATA   -23170,-27246,-30274,-32138
            DATA   -32767,-32138,-30274,-27246
            DATA   -23170,-18205,-12540,-6393
```

Begin main program.

```
            RSXM                          :ensure no sign extension
START
            LAC    COUNT                  :load accumulator
            ADD    STEP                   :update count
            SACL   COUNT                  :save (and wrap)

            LAC    COUNT,5                :isolate integer portion
            SACH   ENTRY                  :save table entry offset

            LACK   STABL                  :point to lookup table
            ADD    ENTRY                  :add offset
            TBLR   SIN                    :read out sine value

            OUT    SIN,DAC                :output to DAC

            B      START                  :back for next
```

PROGRAM DETAILS:

Instruction cycles	14 cycles
Program memory words	11 for code (without initialisation) 32 for table
Data memory words	4 words

Memory reduction techniques To reduce the memory requirement for high accuracy waveform generation, advantage can be taken of waveform symmetry which in effect results in a duplication of stored values. For the case of a sinusoid, the values are repeated four times every period (apart from a sign change), and thus only a quarter of the memory needs to be used to represent the waveform. The price to be paid however is a greater complexity of algorithm (Example 6.2) in order to keep track of which quadrant of the waveform is to be produced and with what sign. The best compromise will be determined by the premium on memory and execution time for a given application and DSP device.

Frequency control As already outlined, the frequency of oscillation or period of waveform repetition is controlled by varying the STEP size of the address counter. The algorithm can thus readily be used to implement a voltage-controlled oscillator (VCO) by causing the step size to be varied in sympathy with the control voltage. One technique is to ADD the control value, FCONT, to the STEP, giving a frequency of oscillation

$$f = (\text{STEP} + \text{FCONT})/(T_s \times N) \text{ Hz}$$

and a frequency deviation δf of

$$\delta f = \text{FCONT}/(T_s \times N) \text{ Hz}$$

For a control value of zero, the oscillator has a well-defined 'free-running' frequency f_0 given by

$$f_0 = \text{STEP}/(T_s \times N) \text{ Hz}$$

Phase control The phase of the oscillator can likewise be controlled simply by ADDing a phase offset signal, PCONT, to the look-up table address pointer, the phase shift $\delta\theta$ being given by

$$\delta\theta = \text{PCONT} \times 360/N \text{ degrees}$$

Note that if phase modulation is desired, the previous phase offset must be SUBtracted from the address pointer prior to the next table read. This ensures that the phase shift is always relative to the same fixed frequency reference. If this is not undertaken, frequency modulation results.

Quadrature oscillators Fixed or variable phase shift between two oscillators is easily accomplished by having a second address pointer which is offset from the main pointer such that the data samples accessed are delayed or phase shifted by the appropriate amount. This technique is particularly useful for obtaining quadrature sinewaves at very low frequencies.

TMS EXAMPLE No. 6.2

Sinewave generation using a reduced length lookup table, taking advantage of symmetry.

Place the lookup table in low program memory, in Q15 notation.

```
              AORG    >20                   :place after vectors

STABL         DATA    0,6393,12540,18205
              DATA    23170,27246,30274,32138
              DATA    32767
```

Begin main program.

```
              RSXM                          :ensure no sign extension
START
              LAC     COUNT                 :load accumulator
              ADD     STEP                  :update count
              SACL    COUNT                 :save (and wrap)
```

The two MSB's of the count value indicate which quadrant we are dealing with.

```
              LAC     COUNT,2               :isolate quadrant
              SACH    QUAD                  :save it
              SACL    ENTRY                 :save remainder of count

              LAC     ENTRY,3               :isolate integer of entry
              SACH    ENTRY                 :save table entry offset

              LACK    STABL                 :point to lookup table
              BIT     QUAD,15               :test if odd quadrant
              BBZ     EQUAD                 :branch if not

              ADLK    8                     :point to end of table
              SUB     ENTRY                 :work backwards

READ          TBLR    SINE                  :read out sine value

              LAC     SINE                  :load accumulator
              BIT     QUAD,14               :test if second half
              BBZ     FIRST                 :branch if first half

              ZAC                           :zero accumulator
              SUB     SINE                  :set up negate

SAVE          SACL    SINE                  :save sine value

              OUT     SINE,DAC              :output to DAC

              B       START                 :back for next
```

Quadrant subroutines.

```
EQUAD         ADD     ENTRY                 :add offset into table
```

| | B | READ | :rejoin main loop |
| FIRST | B | SAVE | :rejoin main loop |

PROGRAM DETAILS:

Instruction cycles	29 cycles
Program memory words	31 (without initialisation) plus 9 for table
Data memory words	5 words

Linear piecewise approximation The main drawback with the direct look-up table technique is the large waveform inaccuracy resulting when a non-integer STEP size is used. The more samples available, the greater the accuracy that can be achieved, but at the expense of increased memory storage requirements. A solution to the problem is to use an interpolation procedure to determine a better estimate of a waveform value that lies between two stored data sample points. The simplest technique is a straight line interpolation procedure (cf. Chapter 3, curve fitting) whereby the intermediate value between two samples is calculated using the formula for a straight line

$$y = mx + c$$

This method is illustrated in Fig 6.3 for an eight-sample sinusoid.

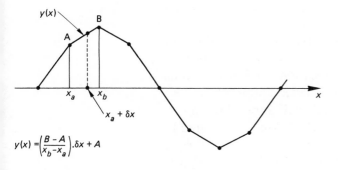

$$y(x) = \left(\frac{B - A}{x_b - x_a}\right).\delta x + A$$

Fig 6.3 Linear piecewise approximation for a sampled sinewave generator

Given that the required waveform value y at time $T_a + \delta t$ lies between two sample values A and B, then y is given by

$$y = [(B - A)/(T_b - T_a)] \times \delta t + A$$

where the time factor $\delta t/(T_b - T_a)$ is simply the fractional part F of the address counter value. Thus, using the direct look-up table algorithm to generate the initial estimate A, all that is required is an additional algorithm which calculates the improvement factor δY given by

$$\delta Y = (B - A) \times F$$

Depending on the constraints on program execution time and memory, the 'slope' term $(B - A)$ can be calculated each program cycle, or stored as predetermined slope values in a second data table.

Example 6.3 gives an implementation using the latter of these two methods and the improved accuracy achieved with this technique can be seen from Table 6.2. Clearly, frequency and phase control can be accomplished in exactly the same manner as for the direct look-up table method.

Improvements can be made in the interpolation technique by using Taylor series or Maclaurin series expansions instead of the straight line approximation, but at the expense of increased algorithm complexity and execution time.

6.2.2 Relaxation oscillators

A second means of waveform generation is the relaxation or counter-based oscillator. The basic element from which more complex waveforms can be derived is the ramp generator (modulo counter, cf. Chapter 3) as shown in Fig 6.4. The waveform has a period P given by

$$P = 2^M T_s / \text{STEP} \quad \text{secs}$$

where M is the modulo counter length, T_s the program execution time, and STEP the counter or ramp increment size.

The ramp waveform as it stands has limited use, due to the very high harmonic content. Ideally, a pure sinusoidal waveform is required for most signal processing applications and hence some means of shaping the ramp waveform to reduce harmonic content is desirable. The accuracy of the sinusoidal approximation is determined largely by the processing time which can be allocated to this process. A simple and time-efficient technique is to synthesize a trapezoidal waveform using the procedure shown in Fig 6.5. This particular waveform has considerably reduced harmonic content compared to the ramp waveform, with all even harmonics suppressed.

The first step in the shaping algorithm is to generate a triangular waveform (usually by rectifying the ramp waveform), and then clipping the peaks to the desired level (cf. Chapter 3, overload conditions). Maximum harmonic suppression is achieved if the

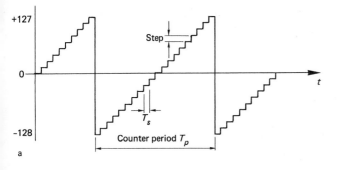

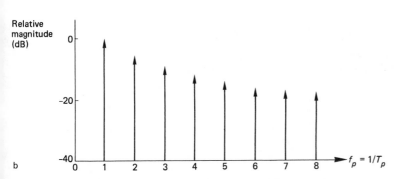

Fig 6.4 Ramp generator:
  a) Time waveform for modulo N counter
  b) Harmonic content of ramp waveform

clipping occurs at two-thirds of the peak value. In this case, the level of the nth odd harmonic relative to the fundamental is given by the relationship:

$$X(n) = \sin(n\pi/3)/(n^2\pi/3)$$

There are a number of ways to implement the waveform-shaping algorithm, one of which is given in Example 6.4.

A second approach to reducing harmonic content is to simply low-pass filter the ramp or subsequent waveforms. However this is in most cases too extravagant in terms of memory usage and execution time.

The frequency of the relaxation oscillator is given by the expression

$$f = \text{STEP}/2^M T_s \quad \text{Hz}$$

Frequency control The frequency can be controlled very simply by altering the STEP size in the ramp counter algorithm. The frequency resolution is again determined by the word length of the STEP register, with a maximum error of

$$\delta f_{max} = 1/2^M T_s \quad \text{Hz}$$

289

TMS EXAMPLE No. 6.3

Sinewave generation using a lookup table (16 values) and interpolation (16 values).

Place the lookup table in low program memory, in Q15 notation. Table holds sine value, followed by difference from next entry.

```
          AORG  >20                    :place after vectors

STABL     DATA  0,12540,12540,10630
          DATA  23170,7104,30274,2493
          DATA  32767,-2493,30274,-7104
          DATA  23170,-10630,12540,-12540
          DATA  0,-12540,-12540,-10630
          DATA  -23170,-7104,-30274,-2493
          DATA  -32767,2493,-30274,7104
          DATA  -23170,10630,-12540,12540
```

Begin main program.

```
START     RSXM                         :ensure no sign extension
          LAC   COUNT                  :load accumulator
          ADD   STEP                   :update count
          SACL  COUNT                  :save (and wrap)

          LAC   COUNT,2                :isolate integer
          SACH  ENTRY,1                :save it
          SUB   ENTRY,15               :leave the fractional part
          SACL  FRACT                  :save it

          LACK  STABL                  :point to lookup table
          ADD   ENTRY,1                :add offset (2 values)
          LARK  AR1,SINE               :set destination (page 0)
          RPTK  1                      :do twice
          TBLR  *+                     :read out SINE, DIFF

          SSXM                         :set sign extension on

          LT    DIFF                   :set up multiply
          MPY   FRACT                  :find inprovement factor
          PAC                          :transfer to accumulator
          ADD   SINE,15                :add in table value
          SACH  SINE,1                 :save final result

          OUT   SINE,DAC               :output to DAC

          B     START                  :back for next
```

PROGRAM DETAILS:

Instruction cycles	26 cycles
Program memory words	22 (without initialisation)
	plus 32 for table
Data memory words	6 words

290

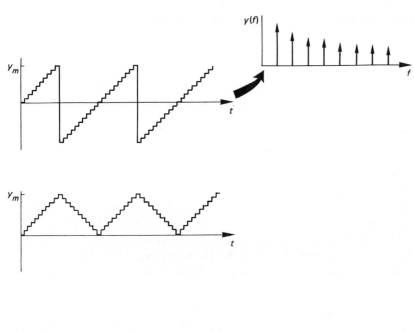

Fig 6.5 Waveform shaping of ramp for reduced harmonic content

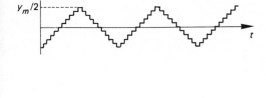

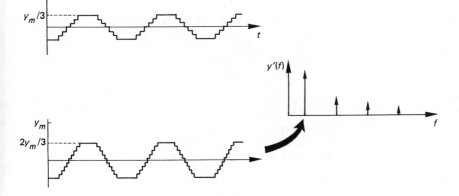

291

If the control value FCONT is ADDed to the STEP value, the VCO frequency is

$$f = (\text{STEP} + \text{FCONT})/2^M T_s \quad \text{Hz}$$

and the deviation

$$f_{dev} = \text{FCONT}/2^M T_s \quad \text{Hz}$$

Phase control The phase of the relaxation oscillator can be controlled by ADDing an offset PCONT to the counter accumulator giving a phase shift $\delta\theta$ of

$$\delta\theta = \text{PCONT} \times 360/2^M \quad \text{degrees}$$

Note that, for phase modulation, the phase offset PCONT must be SUBtracted from the counter accumulator prior to the next calculation so that the phase is always varied relative to a fixed

TMS EXAMPLE No. 6.4

A relaxation oscillator, with waveform shaping.

Begin main program.

	SOVM		:overflow mode on
	SSXM		:sign extension on
START	ZALS	COUNT	:load without sign extension
	ADD	STEP	:update count
	SACL	COUNT	:save (with modulo wrap)

Have produced a sawtooth waveshape - now convert to a triangular wave, and apply shaping.

	ZALH	COUNT	:use high accumulator
	ABS		:produce triangular shape
	SUBH	OFFSET	:(>4000) centre about zero
	SACH	OSC,1	:save result
	ADDH	OSC	:force saturation
	SACH	OSC	:save shaped result
	OUT	OSC,DAC	:output to DAC
	B	START	:back again

PROGRAM DETAILS:

Instruction cycles	13 cycles
Program memory words	12 (without initialisation)
Data memory words	4 words

292

frequency reference (i.e. constant STEP size). If this is omitted, frequency modulation results.

Quadrature signals Precise phase shift between two relaxation oscillators is achieved by adding a fixed offset, as outlined above, to the original counter register, resulting in a second ramp waveform with appropriate phase shift/time delay which can then be shaped as required.

Waveform flexibility Complex waveform shapes are very difficult to synthesize using the relaxation oscillator technique. A small degree of success can be achieved with the use of shaping algorithms, but in general the look-up table method is a much more efficient means of generating non-standard waveforms. Fourier series techniques can be applied to synthesize more complex waveforms, but these result in highly complex and time-intensive algorithms.

6.2.3 Marginally stable recursive filter sections

A third, very useful method of generating sinusoidal waveforms is to use a marginally stable second-order filter section. This recursive oscillator, as it is commonly termed, is in almost all cases the most accurate and efficient method of sinusoidal waveform generation, particularly if quadrature signals are required. A block diagram representation of a recursive oscillator is given in Fig 6.6. The implementation requires just two data memory locations and two multiplications.

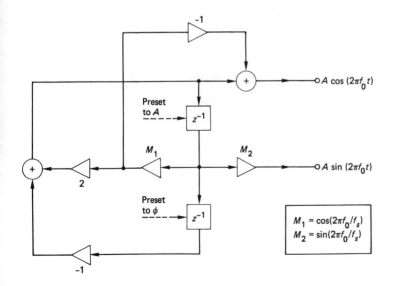

Fig 6.6 Recursive (quadrature) oscillator

293

The frequency of oscillation is determined by the multiplication constant $M1$, with $M2$ acting only as a gain term to ensure equal-level outputs for the cosine and sine waveforms. Clearly, if only a cosine waveform is required, the circuit can be appropriately simplified. The multiplication constant $M1$ is related to the desired oscillator frequency f_0 by the expression

$$M1 = \cos(\omega_0 T_s) = \cos(2\pi f_0/f_s)$$

where f_s is the program sampling frequency. $M2$ is given by

$$M2 = \sin(\omega_0 T_s) = \sin(2\pi f_0/f_s)$$

(Note: for the specific case of $f_0 = f_s/4$, i.e. one quarter of the sampling frequency, then $M1 = 0$ and $M2 = 1$, greatly simplifying the oscillator algorithm.)

To stimulate oscillation in the recursive structure, it is necessary to initialize one of the memory registers with a non-zero value. A simple method is to preset one register to a value A (Fig 6.6) with the other register set to zero. Using this technique, the waveforms generated have a well-defined peak amplitude of A.

The waveform accuracy, in terms of the level of undesired harmonics and frequency resolution, is remarkably good, limited primarily by the processor word length. For example, with 16-bit arithmetic, harmonics are well below 60 dB, and the maximum frequency error δf is approximately

$$\delta f = 1 \times 10^{-3} f_s \quad \text{Hz}$$

(This error reduces significantly for frequencies approaching $f_s/4$.)

An example algorithm for a recursive quadrature oscillator is given in Example 6.5, illustrating the highly efficient implementation that is obtainable both in terms of memory and execution time.

Implementation errors Quantization of the coefficient $M1$ causes the output frequency to be in error by a small amount, whereas quantization of $M2$ causes the quadrature waveform amplitudes to differ slightly. Thirdly, and more seriously, rounding or truncation of the signals generated within the feedback path, i.e. from the $M1$ multiplication stage, can cause the amplitude and phase of the output signal to wander slowly with time.

One particular source of error in this context is signal overload within the loop. Although the level of the output waveform is limited to $\pm A$, the levels within the feedback loop will often be greater. For example, the level prior to the gain stage, $M2$, is greater by a factor of up to $1/\sin(\omega_0 T_s)$. Thus, to be sure of avoiding overload within the oscillator loop, the initialization value should in general be less than the processor overload value by the factor $1/\sin(\omega_0 T_s)$. Gain

TMS EXAMPLE No. 6.5

Use of a marginally stable second order section as an oscillator.

Begin main program.

START	LT	DEL1	:first delay element
	MPY	M1	:gain constant
	PAC		:transfer to accumulator
	SUB	DEL2,15	:second delay element

Note that SUB DEL2,15 is equivalent to multiplying the second delay element by -1.0 and accumulating.

	SACH	COS,1	:save cosine output
	LTD	DEL1	:accumulate, age delay
	SACH	DEL1,1	:update delay
	MPY	M2	:multiply by gain constant
	PAC		:accumulate
	SACH	SIN,1	:save sine output
	OUT	SIN,DAC1	:output sine
	OUT	COS,DAC2	:output cosine
	B	START	:back again

PROGRAM DETAILS:

Instruction cycles	16 cycles
Program memory words	14 (without initialisation)
Data memory words	6 words

compensation can always be applied to the resulting quadrature waveforms to obtain the desired output level.

Correct gain scaling can avoid overload within the feedback loop, but does not eliminate the problem of amplitude and phase wander with time due to truncation or round-off errors. In almost all applications (with word lengths of 16-bits or greater), this effect is negligible and the problem can be ignored. If a problem does arise, then the best solution is to periodically re-initialize the oscillator when the initial states, A and 0, of the registers are repeated.

Frequency control In view of the marginally stable operating condition of the recursive oscillator, any external change of the levels or gain constants within the loop whilst executing is very likely to cause the system to become unstable and continuously variable frequency control is not possible. Discrete frequency changes, however, can be accommodated by re-initializing the oscillator with new gain factors $M1$ and $M2$ as required.

Phase control Again, continuous phase control of recursive oscillators is very difficult, with instability usually arising. Fixed phase offsets between oscillators can be achieved by 'starting' the oscillators at different times, the time difference corresponding to the required phase delay.

6.2.4 Random number/White noise generators

A very important category of waveform generators is random number or random noise generators. Two basic techniques can be used for this purpose, the first being the look-up table method using a random set of stored samples, and the second being based on a shift register with feedback, giving rise to a *pseudo-random binary sequence*. The term pseudo-random is used to describe a sequence which repeats itself after a finite period and is thus not truly random for all time. The sequence length for the look-up table method is quite clearly determined by the number of stored data samples, whilst for the shift register technique it is dictated by the length of the register. This latter technique is considerably more efficient in memory usage than the former and is frequently used for simulating a 'finite band' WHITE NOISE source (i.e. a noise source having a 'flat' or near-constant power spectral density over the frequency band of interest).

Pseudo-random binary sequence (*PRBS*) generators Much in the same way as the second-order recursive filter section with feedback can produce a continuous sinewave, so a shift register with feedback from specific elements can generate a continuous, though repetitive, random sequence of 1s and 0s. A schematic of the system is shown in Fig 6.7. It can be shown [Ref. 3] that the maximum sequence length N_{max} before repetition is given by

$$N_{max} = 2^M - 1$$

where M is the number of stages in the shift register. For example, a shift register of length $M = 8$ gives a sequence length N_{max} of 255.

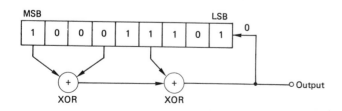

Fig 6.7 Pseudo-random sequence generator

The sequence itself is determined by the position of the feedback taps on the shift register. Only certain feedback taps will result in a maximum length sequence, and these are given in Table 6.3 for shift register lengths M up to 16.

Implementation involves the 'modulo-2' addition (XOR) of the tapped shift register values, yielding either a 1 or a 0, which is then inserted as the LSB of the shift register. The sequence can clearly be output from any of the shift register taps, but in general the result of the modulo-2 addition is used because it is already present in the DSP accumulator. A number of algorithms can be envisaged for PN sequence generation, depending on the length of sequence required.

Register length M	Feedback taps
2	[1,2]
3	[1,3]
4	[1,4]
5	[2,5][2,3,4,5][1,2,4,5]
6	[1,6][1,2,5,6][2,3,5,6]
7	[3,7][1,2,3,7][1,2,4,5,6,7][2,3,4,7] [1,2,3,4,5,7][2,4,6,7][1,7][1,3,6,7] [2,5,6,7]
8	[2,3,4,8][3,5,6,8][1,2,5,6,7,8] [1,3,5,8][2,5,6,8][1,5,6,8] [1,2,3,4,6,8][1,6,7,8]
9	[4,9][3,4,6,9][4,5,8,9] [1,4,8,9][2,3,5,9][1,2,4,5,6,9] [5,6,8,9][1,3,4,6,7,9][2,7,8,9]
10	[3,10][2,3,8,10][3,4,5,6,7,8,9,10] [1,2,3,5,6,10][2,3,6,8,9,10][1,3,4,5,6,7,8,10]
11	[2,11][2,5,8,11][2,3,7,11] [2,3,5,11][2,3,10,11][1,3,8,9,10,11]
12	[1,4,6,12][1,2,5,7,8,9,11,12] [1,3,4,6,8,10,11,12][1,2,5,10,11,12] [2,3,9,12][1,2,4,6,11,12]
13	[1,3,4,13][4,5,7,9,10,13][1,4,7,8,11,13] [1,2,3,6,8,9,10,13][5,6,7,8,12,13][1,5,7,8,9,13]
14	[1,6,10,14][1,3,4,6,7,9,10,14] [4,5,6,7,8,9,12,14][1,6,8,14] [5,6,9,10,11,12,13,14][1,2,3,4,5,7,8,10,13,14]
15	[1,15][1,5,10,15][1,3,12,15] [1,2,4,5,10,15][1,2,6,7,11,15][1,2,3,6,7,15]
16	[1,3,12,16][1,3,6,7,11,12,13,16] [1,2,4,6,8,9,10,11,15,16][1,2,3,5,6,7,10,15,16] [2,3,4,6,7,8,9,16][7,10,12,13,14,16]

Table 6.3 Possible feedback tap positions for maximal-length PN sequence generation [from *Coherent Spread Spectrum Systems* by J. K. Holmes]

Shift register lengths of up to 16 bits can readily be accommodated by a single word on the many 16-bit DSP devices, and thus memory usage is a minimum. Greater-length registers can be implemented by using more than one memory location. An algorithm for generating a 16-bit PN sequence is given in Example 6.6.

White noise generation The average power spectrum of a PN sequence generated using the above technique is illustrated in Fig 6.8. Although not 'flat' across the entire half sampling frequency band, it can for most applications be assumed flat up to frequencies of $f_s/4$, and thus white noise like in nature. (This does not necessarily mean that the signal probability density function is Gaussian in nature.) Clearly, the noise source can be filtered to restrict the band occupancy to within this range if required.

An alternative output from the sequence generator is the entire *M*-bit word (rather than only a single bit). If the contents of the memory register are output after each shift, then the resulting waveform is random in nature but not strictly white noise like (i.e. the signal probability distribution is not truly Gaussian). If the memory register contents are only output after a number of shifts, then the waveform bears a closer resemblance to white noise, becoming 'exact' (within the constraints of a sampled system) when the number of shifts equals or exceeds the shift register length *M*. In other words, to obtain white noise from a 16-bit generator, the memory contents should be output once every 16 shift cycles. In this case, the noise spectrum is flat up to one half of the *waveform output rate* (not the shift rate of the sequence generator).

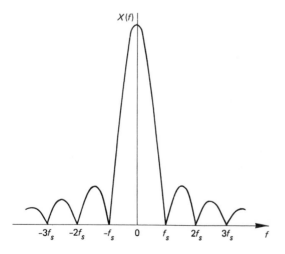

Fig 6.8 Average power spectral density for PN sequence

TMS EXAMPLE No. 6.6

Generation of a pseudo random binary sequence.

A maximal length 16 bit pseudo random sequence is generated, using feedback from bits 0, 2, 11, and 15 (see table 6.3).

Begin main program.

Declare symbols used to reference constants.

```
MASK1    EQU    >8000         :to mask off low accumulator MSB
MASK2    EQU    >8805         :to mask off bits 0, 2, 11 ,15
```

The data memory location holding the sequence should be initialised to a non-zero value.

START

The first method is a general purpose algorithm, to demonstrate the processes involved.

```
LAC     SEQNC,0        :load sequence, align bit 15
ANDK    MASK1          :mask off low accumulator MSB
SACH    TAP1,1         :save bit
```

Note that the ANDK operation clears the high accumulator bits to zero.

```
LAC     SEQNC,4        :load and align bit 11
ANDK    MASK1          :mask off MSB
SACH    TAP2,1         :save bit

LAC     SEQNC,13       :load and align bit 2
ANDK    MASK1          :mask off MSB
SACH    TAP3,1         :save bit

LAC     SEQNC,15       :load and align bit 0
ANDK    MASK1          :mask off MSB
SACH    TAP4,1         :save bit
```

Have separated out the required feedback bits, now combine them to form the 'shift register' input.

```
LAC     SEQNC,1        :load and shift sequence
XOR     TAP1           :use LSB for bit operations
XOR     TAP2           :combine feedback bits
XOR     TAP3           :repeat
XOR     TAP4           :repeat
SACL    SEQNC          :save result (and lose MSB)
```

The required output bit (or word) may now be extracted from the sequence value.

The second method is an optimised routine, which makes use of modulo 2 addition to perform XOR bit operations. It can be applied to this example, but may not be used with certain tap combinations.

```
        LAC    SEQNC              :load sequence
        ANDK   MASK2              :mask off feedback bits
        SACL   SPARE              :save temporarily
```

Combine the feedback bits using the MSB of the accumulator.

```
        ADD    SPARE,4            :combine bits 11 and 15
        ADD    SPARE,13           :combine bit 2 with result
        ADD    SPARE,15           :combine bit 0 with result

        ANDK   MASK2              :re-use mask to mask off MSB
```

If the input bit to the 'shift register' is to form the output, then it is available at this point as the MSB of the low accumulator, and may be saved as appropriate.

```
        ADDH   SEQNC              :combine MSB with sequence
        SACH   SEQNC,1            :save result (and shift out MSB)
```

PROGRAM DETAILS:

Instruction cycles	22 cycles for first method
	11 cycles for second method
Program memory words	22 for first method
	11 for second method
	(without initialisation)
Data memory words	5 words for first method
	2 words for second method

COMMENTS.

The use of data memory to store the mask values would reduce the execution times to 18 cycles for the first method and 9 cycles for the second method.

6.3 Quadrature signal processing

A number of signal processing functions including many described in following sections make use of quadrature processing techniques to improve algorithm efficiency. These techniques are particularly suited to DSP where quadrature waveforms (those which differ by a constant 90° phase shift) can be accurately and effectively realized. A wideband 90° phase shift with constant gain is very difficult to realize in any technology. Usually there has to be a compromise between phase accuracy and gain accuracy, with some techniques providing perfect phase quadrature but poor amplitude matching, and others giving accurate gain but imperfect phase. The degree of gain and phase matching required for satisfactory processing is dependent on the application. Some guidelines are given in the following sections which discuss the use of quadrature processing for specific tasks.

There are three basic algorithms for generating quadrature waveforms: the Hilbert transform filter, the SSB quadrature generator, and time delay (narrowband only).

6.3.1 Hilbert transform filter

The Hilbert transform is the name given to an operator which, when applied to a bandlimited signal, generates the exact quadrature of the waveform (i.e. all components are phase shifted by 90°). In any practical implementation of the transform, the lower the frequency content of the waveform, the greater the difficulty in realizing an accurate quadrature replica. Digital implementation of the Hilbert transform is achieved most readily using finite impulse response (FIR) filtering techniques as described in Chapter 4. These 'filters' have the characteristic of providing an *exact − 90° phase shift* over the entire filter bandwidth, but with a non-constant gain characteristic, the gain accuracy (usually specified as a passband ripple in dB) being determined by the number of stages (taps) used to implement the filter. (Note: the filter gain must always fall to zero at dc.) A typical filter characteristic and implementation is given in Fig 6.9. The filter output is in quadrature with the delayed input signal, the required delay corresponding to the filter delay given by the formula:

$$\text{DELAY} = (N - 1)/2 \text{ samples}$$

The delayed input signal is easily obtained by selecting the appropriate delayed filter input sample. As the filter is *linear phase*, the delay across the band is constant. If an even number of filter taps is used, the delay is a non-integer number of samples and delay matching becomes complicated. (In most cases an additional even-order filter structure must be used as part of the delay element.)

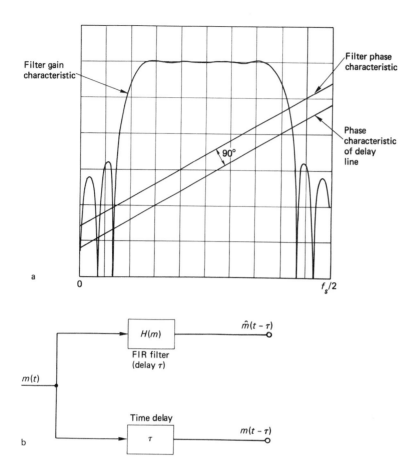

Fig 6.9 Hilbert transform realization:
a) Typical Hilbert transform filter characteristic
b) Implementation of Hilbert transform

Filter gain characteristic

Filter phase characteristic

Phase characteristic of delay line

90°

a

0

$f_s/2$

$H(m)$

$\hat{m}(t-\tau)$

FIR filter (delay τ)

$m(t)$

Time delay

τ

$m(t-\tau)$

b

If it is feasible to choose the sampling frequency and gain characteristics such that the Hilbert transform filter is symmetrical about the quarter sampling frequency point, then all even-filter taps become zero, greatly reducing the algorithm execution time and memory requirement.

A typical Hilbert transform algorithm is given in Example 6.7 for a filter length of 33 stages (taps). The filter characteristic realized is illustrated in Fig 6.9.

Note that, for a filter with an odd number of taps, the Hilbert transform gain characteristic must tend towards zero at dc and at $f_s/2$ (f_s = sampling frequency). For an even number of taps, zero gain at dc is the only constraint.

An alternative to the single filter FIR Hilbert transform implementation is to use two filters, either FIR or IIR, the combined phase of which adds up to 90° (Fig 6.10). The simplest means of designing an FIR filter pair is to apply a quadrature bandpass transform to an appropriate low-pass filter design. The transform involves

TMS EXAMPLE No. 6.7

Hilbert transform, using a Finite Impulse Response filter.

Begin main program.

PG6	EQU	>300	:start of data page 6
PG4	EQU	>200	:page 4 start after CNFD
PG4P	EQU	>FF00	:page 4 after CNFP
NTAPS	EQU	33	:number of filter taps
TDEL	EQU	16	:delay through filter (samples)
START	IN	TAP1,ADC	:input to filter
	CNFP		:block B0 is program memory
	LRLK	AR1,TAP33+PG6	:end tap of filter
	MAC	COF33+PG4P,*-	:perform first multiply
	ZAC		:clear accumulator
	RPTK	NTAPS-2	:repeat for remaining taps
	MACD	COF33+1+PG4P,*-	:multiply, accumulate and age
	APAC		:accumulate final multiply
	SACH	HSIG,1	:save filtered result

The filtered result will be in quadrature with the unfiltered signal, remembering to account for the time delay through the filter.

	OUT	TAP1+TDEL,DAC1	:output unfiltered signal
	OUT	HSIG,DAC2	:output filtered result
	B	START	:back around

PROGRAM DETAILS:

Instruction cycles	52 cycles
Program memory words	16 (without initialisation) plus data tables for constants
Data memory words	67 words

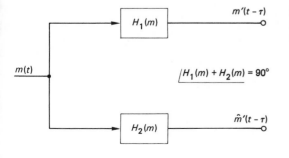

Fig 6.10 Hilbert transform realization using a quadrature filter pair

$m'(t - \tau)$

$\lfloor H_1(m) + H_2(m) = 90°$

$\hat{m}'(t - \tau)$

$H_1(m)$

$H_2(m)$

$m(t)$

multiplying each tap weight, $h(n)$, of the low-pass filter by a cosine operator to yield the taps, $h_1(n)$, of one of the filter pair, and by a sine operator to yield the taps, $h_2(n)$, of the second, where

$$h_1(n) = 2 \cdot h(n) \cdot \cos(2\pi n f_0/f_s)$$

$$h_2(n) = 2 \cdot h(n) \cdot \sin(2\pi n f_0/f_s)$$

This transform yields a filter pair which has a bandpass characteristic centred on the frequency f_0, and twice the width of the corresponding low-pass design (Fig 6.11). The main advantage of this technique over the single filter Hilbert transform is that the two filters have almost identical passband ripple characteristics. This property is of considerable advantage when performing certain of the quadrature processing algorithms described in section 6.6. The most obvious disadvantage is the need to implement two filters rather than one. Note: *only odd-order filters can be designed using this transform.*

An IIR filter pair can result in a much more efficient algorithm in keeping with the advantages of IIR vs. FIR filters in general. The trade-off, as always, however is the non-linear phase response across the filter passband even though the relative phase relationship between the two output signals may be quadrature. The quadrature transform design method does not apply in this case.

Fig 6.11 Realization of Hilbert transform using SIN/COS transformation

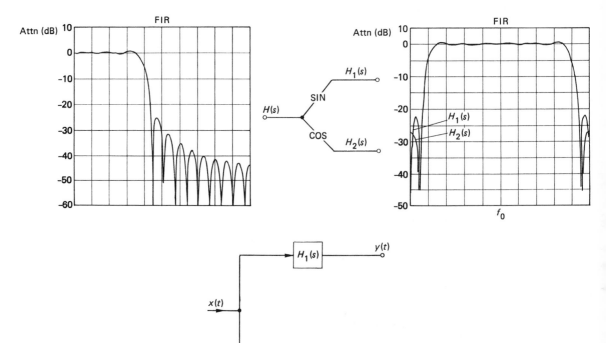

6.3.2 SSB quadrature generator

A second method of quadrature waveform generation is to use a process of double balanced modulation (cf. section 6.5) and single sideband filtering. This process is illustrated schematically in Fig 6.12. By multiplying the input waveform with quadrature carrier signals, two sets of quadrature 'sidebands' are generated. The lower or upper sidebands in each sideband pair is suppressed by filtering, leaving two frequency translated versions of the original waveform having the desired quadrature relationship. As the frequency of the modulating waveform approaches zero hertz, the sideband pair become closer together and eventually cannot be separated by 'practical filtering', the complexity and order of the sideband filter increasing as the input waveform frequency relative to the sampling frequency is reduced. This factor imposes a practical limit on the frequency band over which the quadrature relationship can be maintained.

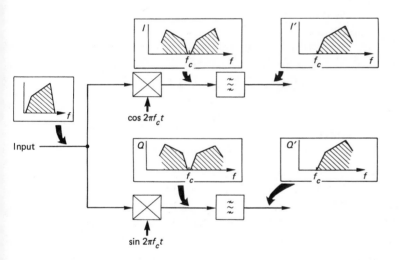

Fig 6.12 Quadrature frequency translation using SSB modulation

A particularly useful form of the SSB quadrature generator is known as the *weaver method*, which is characterised by the carrier frequency being chosen to correspond to the centre frequency of the modulating signal. This has the effect that the resulting lower sideband is folded about zero hertz (Fig 6.13) and thus occupies the lowest frequency range possible. The sideband selection filter is now a simple low-pass filter, and in many cases can be configured as a *halfband filter* for optimal algorithm efficiency (cf. Chapter 4).

The two sidebands generated by the above technique are necessarily translated in frequency, requiring a second stage of mixing to restore the quadrature waveforms to the original frequency segment.

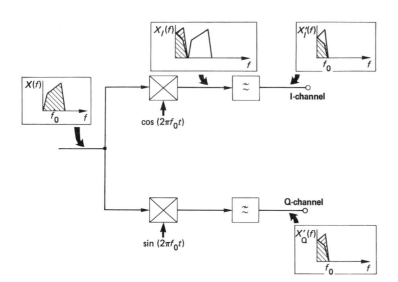

Fig 6.13 Weaver-based quadrature frequency generation

If a quadrature frequency shift procedure is used for this purpose (section 6.6) then no further filtering is required.

In many of the processing techniques described in following sections, this second stage of mixing and filtering is superfluous. (Note: care must be taken when implementing any frequency translation process to take account of aliasing products, cf. Chapter 2.)

The SSB quadrature waveform generation technique is seemingly very much more complex and less efficient than the Hilbert transform filter technique, but has two distinct advantages. Firstly, the two sideband filters can be made identical, saving on memory requirements and imposing identical phase and gain characteristics on both bands. This latter facility is extremely important for certain quadrature processing applications, namely frequency translation discussed in a subsequent section. The second advantage in some applications is the inherent frequency translation involved.

6.3.3 Narrowband phase shift

The very simplest form of phase shifting for a fixed frequency signal is to delay the signal by an appropriate number of samples. Clearly, for frequencies well below the sampling rate of the processor, a large amount of memory is required to achieve large phase shifts, and the technique is less attractive. The accuracy with which $90°$ phase shifts can be implemented is dictated by whether the delay required corresponds to a whole number of program cycles. In other words, the number of samples in one period of the waveform must be exactly divisible by four. Although only precise for a single frequency, the phase shift obtained using this method will be reasonably accurate over a small band of frequencies about the chosen component and the technique can be applied very effectively for quadrature waveform generation in these cases. The reader is referred to the section on delay in Chapter 3 for methods of algorithm implementation.

6.4 Signal detection

This section presents a number of algorithms for the detection of signal envelope, frequency and phase. Any waveform $y(t)$ can be represented as an envelope and phase modulated sinusoid of the form

$$y(t) = r(t) \cdot \cos[\omega_c t + \phi(t)]$$

where $r(t)$ represents the time-varying envelope of the waveform and $\phi(t)$ the time-varying phase. The time derivative of $\phi(t)$, denoted $\dot{\phi}(t)$, is referred to as the instantaneous frequency deviation of the waveform.

6.4.1 Envelope detection

This section presents a number of algorithms for envelope detection. It is by no means exhaustive but includes most of the methods applicable to DSP implementation.

Rectification and filtering One of the most straightforward envelope detection algorithms involves rectification of the input signal followed by a low-pass filter. The rectification can either be implemented as half-wave or full-wave rectification, the two types being illustrated in Fig 6.14.

In the vast majority of cases, full-wave rectification, ABS(X), is the most applicable, giving a greater power in the recovered envelope component than half-wave rectification and with the fundamental input frequency suppressed.

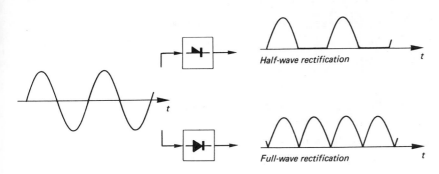

Half-wave rectification

Full-wave rectification

Fig 6.14 Examples of half-wave and full-wave rectification

307

If we consider a modulated sinusoid given by

$$y(t) = r(t) \cdot \cos[\omega_c t]$$

where $r(t)$ is the signal envelope (the phase modulation $\phi(t)$ has been dropped for convenience), then the Fourier series of the half-wave and full-wave rectified signal is

$$y(t)_{\text{half-wave}} = \frac{1}{\pi} + \frac{1}{2}\sin\omega_c t - \frac{2}{\pi}\left[\frac{1}{3}\cos 2\omega_c t + \frac{1}{3.5}\cos 4\omega_c t + \frac{1}{5.7}\cos 6\omega_c t + \cdots\right]$$

$$y(t)_{\text{full-wave}} = \frac{2}{\pi} - \frac{4}{\pi}\left[\frac{1}{3}\cos 2\omega_c t + \frac{1}{3.5}\cos 4\omega_c t + \frac{1}{5.7}\cos 6\omega_c t + \cdots\right]$$

Thus, if the input signal $y(t)$ has the spectrum shown in Fig 6.15a, then the spectrum of the full-wave rectified signal is that of Fig 6.15b. For both half-wave and full-wave rectification, infinite even-order harmonics of the input signal are generated in addition to the desired envelope term. The subsequent filtering must therefore be sufficient to remove the lowest of the unwanted harmonics whilst passing the envelope terms without distortion. In some cases the envelope components and signal harmonics overlap in frequency and it is impossible to unambiguously extract the desired envelope waveform. Under these circumstances, one of the alternative envelope detection techniques presented below can be applied.

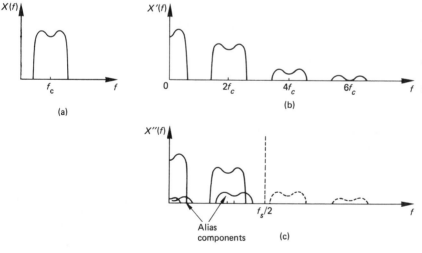

Fig 6.15 Spectra resulting from full-wave rectification:
a) Input spectrum
b) Full-wave rectified signal spectrum
c) Aliased signal spectrum

In a sampled system, the non-linear rectification process can create significant problems, with the resulting higher-order harmonics giving rise to alias components folding back into the frequency band of interest (Fig 6.15c). By careful choice of sampling frequency, most of the significant alias components can be made to fall outside the desired frequency band or are at least sufficiently attenuated to cause little distortion to the envelope term.

Squaring envelope detection Most of the problems of aliasing associated with waveform rectification can be eliminated by squaring the signal to obtain an envelope term. The resulting signal components are given by

$$[y(t)]^2 = \tfrac{1}{2} r(t)^2 + \tfrac{1}{2} r(t)^2 \cdot \cos[2\omega_c t + 2\phi(t)]$$

which include a term proportional to the square of the envelope.

Whilst eliminating the large number of harmonics resulting from signal rectification (only the twice-frequency term remaining), the squaring technique has two major drawbacks. Firstly, the sampling rate must still remain high to ensure that the $2f$ term generated does not cause aliasing. Secondly, the square of the envelope is generated rather than the envelope itself, necessitating the use of square-rooting algorithms (cf. section 3.5.3) to obtain $r(t)$. The squaring process also has the unfortunate property of doubling the signal dynamic range and hence the number of bits required to represent the squared envelope compared with the envelope term itself. This constraint in itself can make square-rooting algorithms difficult to perform without recourse to floating-point arithmetic.

In some applications—for example the detection of the envelope of a two-tone signal—the envelope squared term may in fact occupy a much narrower bandwidth than the envelope itself, making extraction easier in the final filter stage.

Quadrature envelope detection Quadrature envelope detection (Fig 6.16) exploits the trigonometric relationship

$$[r(t) \cdot \cos(\omega_c t)]^2 + [r(t) \cdot \sin(\omega_c t)]^2 = r(t)^2 \cdot 1$$

which can be seen to yield an envelope squared component as for the squaring detector but without the $2f$ term. Consequently, no further

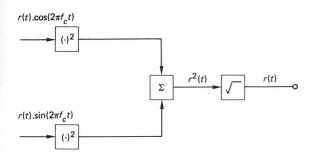

Fig 6.16 Quadrature envelope detection process

filtering is required and the problems of spectral overlap are eliminated. If FIR-based Hilbert transform filter techniques are used to generate the initial quadrature terms (cf. section 6.3), then *minimal sampling rate* algorithms can be implemented as all alias components cancel out. The major drawbacks are the increased algorithm complexity to generate the quadrature signals and cope with the ensuing squared envelope term as mentioned above. If quadrature versions of the waveform are already available, then this technique is very attractive.

Peak signal envelope detection An approximate method of envelope extraction is to simply detect the peak positive (or negative) values of a waveform as illustrated in Fig 6.17. The accuracy of this method is determined by the accuracy with which the peak of the waveform value can be found and the ratio of carrier frequency to the envelope modulation frequency. Clearly, the more samples per waveform period examined, the greater the accuracy that can be achieved. Envelope distortion increases dramatically as the signal frequency approaches the sampling rate used.

Fig 6.17 Peak signal envelope detection

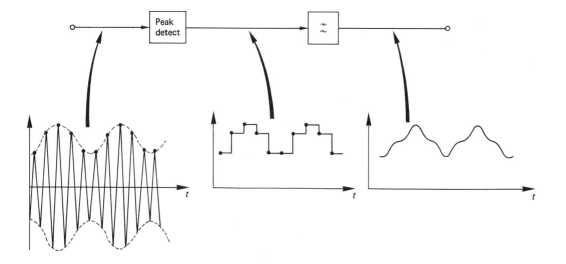

The attraction of this particular algorithm is that no $2f$ terms or subsequent harmonics are generated in the detection process and the approximation to the envelope term is derived directly (not the envelope squared). Also, in many applications, no filtering is required to extract the envelope other than a possible waveform smoothing filter as shown in Fig 6.17.

The actual peak signal detection algorithm can be implemented in a number of ways. Comparison of present and previous stored data values to detect a peak level during one waveform period is the simplest solution. This technique however requires the monitoring of the waveform zero crossings to determine the start and end of a period of observation. The major drawback of the technique is the sensitivity to noise which can cause large false peaks in the waveform characteristics.

6.4.2 Frequency detection

This section explores a number of algorithms which perform frequency detection, often referred to as frequency discrimination or frequency counting. An important consideration in the selection of an algorithm is whether the information required is to be simply a measure of the *relative* frequency of a signal, or the *absolute* frequency of a signal measured in hertz or radians/second. The former case, satisfactory for most applications, is usually referred to as *frequency detection or discrimination*, whilst the latter case is often termed *frequency counting*. This distinction between absolute and relative frequency measurement will be made clear in the following algorithm descriptions. A further distinction must be made between *instantaneous frequency*, defined as the time derivative of the waveform phase variations, and the *periodic frequency* of a waveform.

To aid the correct choice of algorithm for a given application, the properties of each technique are summarised in Table 6.4.

Detection method	Property: Relative frequency	Absolute frequency	Instantaneous frequency	Algorithm complexity
Filter bank	YES	NO	NO	LOW
Filter slope	YES	YES	YES	LOW
PLL	YES	YES	YES	HIGH
Pulse counting	YES	YES	NO	LOW
Quadrature	YES	YES	YES	MEDIUM
FFT	YES	YES	NO	HIGH

Table 6.4 Measurement capabilities of various frequency detectors

Filter-based frequency detectors If the requirement of the algorithm is simply to detect the presence or absence of one or more closely spaced frequency components, then a bandpass filter followed by an envelope detector is usually adequate for this purpose. This technique simply provides an indication of the presence of a frequency component within the appropriate filter passband (Fig 6.18). The viability of the technique is governed by the selectivity of the bandpass filters which dictates the minimum

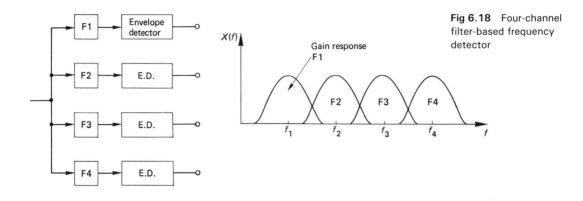

Fig 6.18 Four-channel filter-based frequency detector

spacing between sets of frequency components, before reliable discrimination is lost. In most applications, IIR filter realizations are a perfectly adequate and efficient solution. This technique is frequently used in the detection of binary Frequency Shift Keyed signalling where the receiver needs to distinguish between two well-spaced components representing the transmission of a logic 0 or logic 1.

Filter slope frequency discriminator A second filter-based frequency discriminator makes use of the passband to stopband transition to provide a well-defined frequency to voltage conversion characteristic. Provided that the input signal frequency range falls within the filter transition region then the output level from an envelope detector following the filter will be directly proportional to the position of a component within the transition region and hence its frequency. This property is illustrated in Fig 6.19. The use of low-pass or high-pass transitions will provide an output level which decreases or increases with frequency respectively. If the input level to the filter discriminator is known, then the signal frequency can be accurately determined from a knowledge of the filter transition characteristics and the envelope detector gain constant. If input

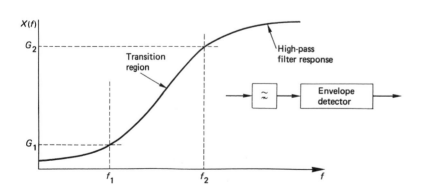

Fig 6.19 Filter-based frequency to voltage converter

312

signal level is unknown, then only a relative frequency estimate can be obtained.

The main advantage of this technique over the bandpass filter detector is that frequencies can be discriminated over a continuous range (i.e. the filter transition region), rather than at discrete values coinciding with filter passbands.

The drawback of filter discriminators is the restricted linearity, range and resolution which can realistically be achieved.

Phase locked loop discriminators The phase locked loop discriminator can be viewed as either a type of discrete bandpass filter signal detector or a continuous frequency discriminator, depending on the processing involved. Detailed description of PLL operation can be found in Ref. 6.2 and a limited knowledge is assumed in this discussion.

The main advantage of using a PLL as a bandpass filter is that it can be made extremely selective by using narrow loop bandwidths. If a signal component falls within the *capture range* of a PLL then the loop will lock onto that component and a lock indicator used as the detector output signal.

By monitoring the VCO control voltage of a locked PLL, a continuous frequency discriminator can be realized. The loop control voltage is directly proportional to the VCO frequency (assuming a linear voltage to frequency conversion characteristic), and hence is also proportional to the input signal frequency, when the loop is locked. A suitably scaled control voltage waveform thus provides a measure of the instantaneous frequency variations of the input waveform provided that the PLL can track the input signal (Fig 6.20).

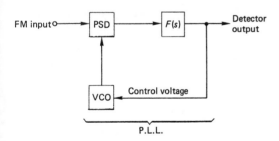

Fig 6.20 Phase locked loop (PLL) frequency discriminator

Both applications of the PLL for frequency detection are frequently used in analog signal processing. However, for DSP implementation, a number of the following techniques are generally more efficient and accurate.

Pulse counting/averaging The pulse-counting technique simply involves assessing the number or rate of zero crossings of a waveform within a given period of time. This can be used as either a relative measure of frequency or an absolute frequency counter if the measurement period is precisely defined. (Note: this technique is only valid when the waveform has a well-defined number of zero crossings per period.) The zero crossing frequency discriminator can be realized in several ways using DSP techniques, two of which are illustrated in Figs 6.21 and 6.22. Fig 6.21 depicts the conventional 'analog' technique whereby fixed-width pulses are generated at each zero crossing and then filtered to determine the average power density. The higher the signal frequency, the closer the zero crossings and hence the pulse spacing, giving a high-level output, whereas the lower the frequency, the fewer zero crossings and the lower the output level. The same technique can be used as the basis for a DSP frequency detection algorithm.

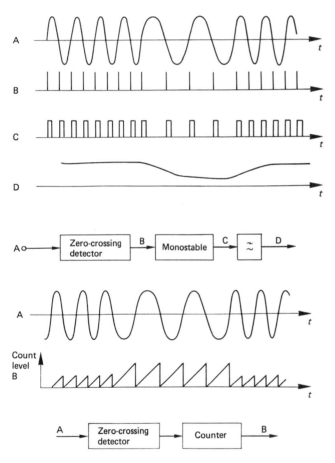

Fig 6.21 Pulse position discriminator

Fig 6.22 Pulse counting discriminator

A more efficient technique is to actually count the time between zero crossings to give a direct measure of frequency. This method is illustrated in Fig 6.22. A zero crossing of the input waveform is detected by testing the signal sign bit. When a zero crossing occurs, a counter is started which increments each program cycle until the next zero crossing is encountered. At this point, the count value is stored and the counter reset. A direct measure of the time between zero crossings can therefore be obtained, and hence the fundamental waveform frequency. The counter output values can be filtered for further signal averaging if required, although this is not strictly necessary. This algorithm has applications for both relative and absolute frequency measurement since the frequency is given exactly by the equation:

$$\text{Freq} = 1/\text{Period} = 1/[2 \cdot \text{COUNT} \cdot T_s]$$

where T_s is the program cycle time or sampling period.

The accuracy of the frequency measurement is determined directly by the sampling rate or program cycle time which sets the time resolution with which zero crossings are detected. Better accuracy can usually be achieved by counting over a number of zero crossings and taking the average value. (This is equivalent to using a longer gate period in conventional frequency counters.) An example algorithm for a simple zero-crossing frequency detector is given in Example 6.8.

Quadrature frequency detectors   Although the counting techniques described above can give an efficient measure of absolute and relative periodic frequency, they do not detect the instantaneous frequency deviations of a waveform which result from continuously varying phase slope between zero crossings. This measurement can be made by the PLL technique already described or more efficiently by the use of quadrature processing techniques.

The quadrature frequency detection algorithm is illustrated in Fig 6.23. If the input signals are of the form

$$I(t) = \tfrac{1}{2} r(t) \cdot \sin \phi(t)$$

and

$$Q(t) = \tfrac{1}{2} r(t) \cdot \cos \phi(t)$$

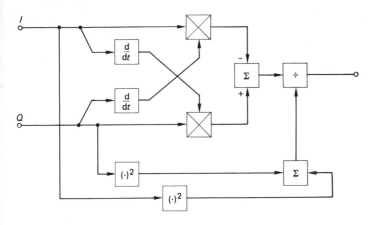

Fig 6.23 Quadrature frequency discriminator

TMS EXAMPLE No. 6.8

Simple, zero crossing frequency detector.

Begin main program.

	SSXM		:sign extension on
START	IN	INPUT,ADC	:input signal
	LAC	INPUT	:load with sign extension
	SACH	SGN1	:save high accumulator

Due to sign extension, the high accumulator holds >0000 for positive signals or >FFFF for negative ones.

	ZALS	SGN2	:load previous sign
	DMOV	SGN1	:age current sign
	XOR	SGN1	:perform comparison
	BZ	NOCROSS	:branch if signs are the same

The result of the XOR will be zero if SGN1 and SGN2 are the same, indicating no zero crossing.

	DMOV	COUNT1	:store current count
	ZAC		:clear accumulator
	NOP		:padding
	NOP		: "
SAVCNT	SACL	COUNT1	:save updated or reset count
	OUT	COUNT2,DAC	:output count to last crossing
	B	START	:back again

No zero crossing branch

NOCROSS	LAC	COUNT1	:load current count
	ADD	ONE	:update it
	B	SAVCNT	:rejoin main loop

PROGRAM DETAILS:

Instruction cycles	18 cycles
Program memory words	20 (without initialisation)
Data memory words	5 words

where $r(t)$ is the signal envelope and $\phi(t)$ the angular phase/frequency, then the signals at the outputs of the two differentiators can be represented as

$$\dot{I}(t) = \tfrac{1}{2} r(t) \cdot \dot{\phi}(t) \cdot \cos \phi(t) + \tfrac{1}{2} \dot{r}(t) \cdot \sin \phi(t)$$

and

$$\dot{Q}(t) = -\tfrac{1}{2} r(t) \cdot \dot{\phi}(t) \cdot \sin \phi(t) + \tfrac{1}{2} \dot{r}(t) \cdot \cos \phi(t)$$

By cross-multiplying and subtracting these signals as shown, an output signal is obtained given by

$$\dot{I}(t) \cdot Q(t) - I(t) \cdot \dot{Q}(t) = \tfrac{1}{4} r^2(t) \cdot \dot{\phi}(t)$$

This signal is directly proportional to the rate of change of phase or instantaneous frequency of the input signal and also proportional to the envelope squared. In order to remove the envelope variation, the signal output must be divided by the squared envelope itself, derived using the quadrature technique described in section 6.4.

Although an apparently complex algorithm for frequency detection, most of the elements are very simple to realize. The differentiation for example can be performed using a simple FIR filter as discussed in Chapter 4. The major attraction of this technique is that the frequency detection is continuous (i.e. a true measure of the rate of change of phase of the waveform). The major drawback is the time-consuming envelope normalization process. This latter constraint can be eliminated if the envelope variations are removed prior to the generation of quadrature components.

If suitably calibrated, the output from the quadrature detector can be used to give an accurate measure of absolute frequency as well as relative frequency.

Fast Fourier Transform (FFT) Chapter 5 describes and analyses algorithms for obtaining the FFT of a time waveform. Discrete frequency components within a waveform can be isolated using this technique, the resolution depending on the number of waveform samples processed by the FFT algorithm. In this respect, the FFT detection process is similar in nature to a large bank of bandpass filter frequency detectors. FFT techniques are by their very nature discrete in operation, i.e. they give only a 'snapshot' of the frequency components constituting a signal in a given period of time, and thus provide no useful measure of the instantaneous frequency variations of the waveform. Secondly, they require complex time-consuming algorithms, particularly if high resolution is required. (Careful choice of DSP device will help minimize the processing time for FFT operations!)

6.4.3 Phase detection

Following on from frequency detection this section deals with methods and algorithms for phase detection.

The first point to stress about phase detection is that phase is always measured relative to a reference, which, if known, permits absolute phase measurement and, if not known, permits only relative phase measurement.

The phase of a signal can theoretically be derived from the instantaneous frequency by simply integrating the frequency term. The problem with this approach however is that the 'constant of integration' is not usually known, and so only a relative measure of phase can be obtained using this technique. This method is not pursued further here in view of the implementation complexity compared with alternative phase detection algorithms presented below.

Mixer-based phase detector   Phase detectors based on a mixing operation can be used to determine the phase difference between two signals. Suppose we wish to determine the phase $\phi(t)$ of a signal $y(t)$ given by

$$y(t) = \cos[\omega_c t + \phi(t)]$$

relative to a reference term $z(t)$ given by

$$z(t) = \sin[\omega_c t]$$

Mixing the two signals generates the terms

$$y(t) \cdot z(t) = \sin[\phi(t)] + \sin[2\omega_c t + \phi(t)]$$

Ignoring the $2f$ term, the required phase term $\phi(t)$ is obtained in the form $\sin[\phi(t)]$. If $\phi(t)$ is small, then the approximation $\phi(t) = \sin[\phi(t)]$ can be used. However, if $\phi(t) > 20°$, then a curve fit or expansion routine for $\sin^{-1}[\phi(t)]$ must be used to accurately determine $\phi(t)$. Note: $\phi(t)$ can only be determined without ambiguity if $\phi(t) < \pm 90°$. ($\sin 91° \equiv \sin 89°$.)

A similar analysis can be performed assuming a cosine reference $z(t)$ given by

$$z(t) = \cos[\omega_c t]$$

In this case, the mixer output becomes

$$y(t) \cdot z(t) = \cos[\phi(t)] + \cos[2\omega_c t + \phi(t)]$$

The desired phase term is now in the form $\cos[\phi(t)]$ which means that provided the phase is known to be lagging or leading the reference, a phase difference of up to $180°$ can be detected without

318

ambiguity. (Note: it is no longer possible to distinguish between positive and negative phase shifts with the COS function.)

An example of a $\sin^{-1}[\phi(t)]$ curve fit algorithm is given in Example 6.9, valid for $\phi(t) < \pm90°$, together with a data table for implementing $\cos^{-1}[\phi(t)]$.

If only discrete phase shifts are to be detected, as in the case of phase shift keyed data modulation, then the $\sin[\phi(t)]$ (or $\cos[\phi(t)]$) phase term can be used directly as an indication of the received signal phase.

The mixing technique has the attraction of giving a continuous and absolute measure of phase (over a limited range), but suffers from the need for a fairly complex curve-fitting algorithm for phase errors in excess of about 20°. This drawback can be overcome by using counting techniques.

Counting phase detectors An alternative phase detection technique (Fig 6.24) is based on the zero-crossing detector and counter principle used for frequency detection. By measuring the time which elapses between a zero-crossing of a reference and a corresponding zero-crossing of the unknown waveform, the phase difference can be evaluated using the simple relationship

PHASE = TIME ELAPSED · 360°/PERIOD

assuming the waveform period is known.

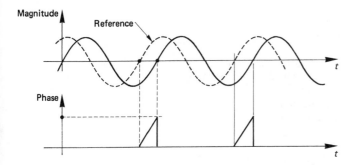

Fig 6.24 Pulse counting phase detector

This counting method can determine phase shifts within $\pm180°$ without ambiguity, and up to 360° if the waveform is known to be either lagging or leading the reference.

The accuracy of phase measurement is determined by how accurately the waveform zero-crossings can be detected, and how precise the elapsed time and period measurement is. All these factors are improved as the number of samples processed per waveform period is increased.

TMS EXAMPLE No. 6.9

Evaluating sin^{-1} by means of a linear curve fit.

Begin main program.

MSB	EQU	0	:for bit testing

Place curve-fit lookup table in program memory. Ten linear ranges, end-point and difference from next end-point stored for each range.

	AORG	>20	:place after vectors
CFTBL	DATA	0,2090,2090,2010	
	DATA	4200,2156,6356,2229	
	DATA	8585,2338,10923,2501	
	DATA	13424,2751,16175,3169	
	DATA	19344,4015,23359,9408	

	SOVM		:overflow mode on
START			
	ZALH	SIN	:load sine value
	ABS		:extract magnitude
	SACH	ENTRY	:save it
	LAC	ENTRY,1	:load 2*magnitude
	ADD	ENTRY,3	:accumulator holds 10*magnitude
	SACH	ENTRY,1	:save integer portion

ENTRY will hold a value between 0 and 9 inclusive, indicating the curve-fit range.

	SUB	ENTRY,15	:remove integer
	SACL	FRACT	:save fractional part
	LALK	CFTBL	:point to lookup table
	ADD	ENTRY,1	:add offset (2 values per range)
	TBLR	ISIN	:read end-point
	ADD	ONE	:update source
	TBLR	DIFF	:read difference
	LT	FRACT	:set up multiply
	MPY	DIFF	:calculate improvement factor
	PAC		:transfer to accumulator
	ADD	ISIN,15	:produce interpolated result

Now restore the sign of the result. Note that the corresponding cos^{-1} curve-fit would not require this stage.

	BIT	SIN,MSB	:check sign of sine
	BBZ	POSITV	:branch if positive
	NEG		:otherwise, negate accumulator

```
         NOP                          :padding

SAVRES   SACH   ISIN,1                :save final result

         B      START                 :back around
```

Positive branch.

```
POSITV   B      SAVRES                :branch around negation
```

Cos^{-1} *curve fit table.*

```
CFTBL    DATA   32767,-2089,30678,-2110
         DATA   28568,-2156,26412,-2229
         DATA   24183,-2338,21845,-2501
         DATA   19344,-2751,16593,-3169
         DATA   13424,-4015,9409,-9409
```

PROGRAM DETAILS:

Instruction cycles	30 cycles
Program memory words	28 (without initialisation) plus 20 for lookup table
Data memory words	6 words

The disadvantage of the counting technique compared with the mixer method is that only discrete phase measurements can be made, i.e. once every reference zero-crossing.

6.5 Modulation techniques

Modulation is the term given to the process of varying one or more properties of a well-defined reference or carrier signal, according to the nature of a second message signal, in order to convey information in a manner suited to the transmission medium.

This section looks at the DSP realization of Amplitude Modulation (AM), Phase Modulation (PM), and Frequency Modulation (FM). *In all modulation operations, the resultant modulated signal consists of frequency components not present within either of the generating signals, and considerable care must be exercised when designing DSP modulators with regard to aliasing.* This is particularly true of FM or PM systems as illustrated in Fig 6.25. Reference is made to much of the material given in section 6.2 on waveform generation.

Fig **6.25** Typical spectra
for AM, FM and PM signals
illustrating effects of aliasing

6.5.1 Amplitude modulation

Double balanced mixers The term 'double balanced mixer' arises
from analog signal processing where careful design is required to
obtain a mixer which generates the product of the two input
waveforms without any residual (feedthrough) components from the
mixer inputs. This is achieved in practice by carefully balancing the
circuitry used such that undesired feedthrough terms cancel out. The
advantage of DSP multipliers in this context is that precise binary
arithmetic is used, ensuring a perfect multiplication process with no
feedthrough.

To generate a double sideband suppressed carrier AM (DSB)
signal, only one multiplication operation is required. If full carrier
AM is to be realized, then the carrier component must be added back
in after the mixer stage. (An alternative technique is to add a
dc component to the message signal prior to mixing which results
in a residual carrier component in the modulated signal.) These
techniques are illustrated in Fig 6.26.

Note: dc offsets on the input waveform, arising from analog
interface circuitry, can often cause undesirable carrier feedthrough
in a mixing process. The best solution is to pass the input signal
through a digital ac-coupling filter (section 3.6) prior to mixing.

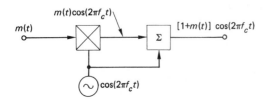

Fig **6.26** Full-carrier
amplitude modulator

Look-up table method   If information is to be conveyed in discrete amplitude level changes, such as data transmission using 4-level ASK, then a look-up table containing the appropriate waveform can be accessed when a particular message bit is to be sent. (The use of a look-up table for waveform generation is discussed more fully in section 6.2.) The look-up table technique is also universally applicable to discrete frequency or phase modulation, or even combinations of all three modulation types. The limitation of the technique, apart from the discrete signalling nature, is the large amount of memory required to store the look-up table constants.

Barrel shifting   A modulation method peculiar to DSP implementation is the use of left or right shift data commands to introduce discrete (powers of two) changes in the binary value of a waveform and hence its equivalent analog amplitude. In some applications, this can lead to a more efficient algorithm than conventional multiplication/mixing by the appropriate constant.

6.5.2   Frequency modulation

Voltage-controlled oscillator   The use of a VCO for frequency modulation is probably the most widely adopted analog technique and is also one of the most efficient DSP implementations. Various realizations of VCOs are discussed in section 6.2. The main attraction of the VCO method is that continuous frequency variations can be synthesized, compared with the discrete frequency modulation produced by all other types of FM modulator. This means that continuously varying, or analog, information can be conveyed and detected.

Waveform switching   As for the AM case, discrete frequency modulation can be achieved by selecting a waveform generator with frequency representative of the desired message symbol. Whilst this is a very efficient technique for conveying a small number of message symbols (and hence few waveform generators), the resulting time waveform obtained is discontinuous when switching occurs, unless the waveform sources are harmonically related and phase locked, or the switching time is carefully chosen. The technique is often used for simple frequency shift keyed data transmission.

Look-up table   The look-up table can be used for implementing frequency modulation, acting as a VCO, or as a discrete frequency modulator with different frequency waveforms stored in individual tables. This latter approach is grossly inefficient compared with the VCO method (only one look-up table) if discrete sinusoids are to be generated. However, it provides a very attractive means of generating complex frequency/phase time waveforms for each discrete message symbol which may have useful properties such as low bandwidth occupancy or high interference immunity. In this respect, the look-up table is often the only practical method of implementation.

6.5.3 Phase modulation

Discrete phase modulation A number of discrete phase modulation algorithms have already been discussed in section 6.2 on waveform generation and the reader is referred to this section for further details. One technique not yet covered is the waveform switching method whereby discrete phase states are obtained by switching to an appropriate waveform generator or look-up table operating with the required phase shift. This method is frequently used for digital phase modulation.

Armstrong phase modulator A means of achieving continuous phase modulation is the Armstrong modulator (Fig 6.27) which makes use of the approximation

$$\cos[\omega_c t + \phi(t)] \doteq \cos(\omega_c t) - \phi(t) \cdot \sin(\omega_c t)$$

valid for $\phi(t) < 10°$ without introducing undue distortion. The attraction of this modulation technique is its algorithmic simplicity, involving only one quadrature signal generator and a multiplier stage. The limitation is unfortunately the restricted phase modulation range.

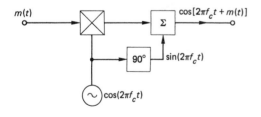

Fig 6.27 Armstrong phase modulator

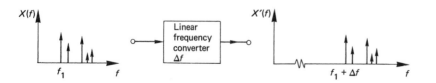

Fig 6.28 Linear frequency translation

6.6 Frequency translation

Before discussing particular algorithms for frequency translation, it is worth defining what we are attempting to achieve. The modulation techniques described in the previous section can be viewed as providing frequency translation, but in every case the modulated signal contains more spectral components than the original message waveform, and in many cases bears no resemblance to the spectrum of the message signal. The frequency translation algorithms to be presented here realize a direct linear frequency translation of all message frequency components without altering the spectral distribution and without creating any additional components. This process is quite often referred to as Single Sideband Modulation (SSB) or true Linear Modulation (LM) and is illustrated in Fig 6.28.

Filtering technique   The filtering technique has already been introduced in section 6.3 on quadrature waveform generation. Upwards or downward frequency translation can be achieved by mixing (multiplying) the input waveform with a local oscillator whose frequency corresponds to the desired frequency shift, and selecting the appropriate upper or lower sideband (Fig 6.29). Provided that the local oscillator frequency is greater than half the highest frequency of the input signal, then complete rejection of the unwanted sideband is possible (assuming ideal brickwall filters and no aliasing). If however the local oscillator frequency is less than half the input frequency, the sidebands partially overlap, and perfect frequency translation is impossible. With practical filters having a finite transition region, the translation of low-frequency components cannot be realized without either undue suppression of the low-frequency terms in the wanted sideband, or poor rejection of low-frequency terms in the unwanted sideband. This is the major drawback of the filter technique for frequency translation, coupled with the restriction on the use of low-frequency local oscillators, and hence small frequency shifts, due to sideband overlap.

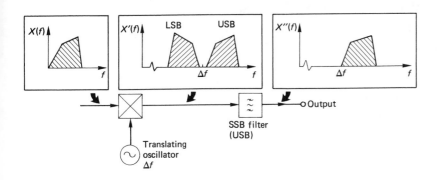

Fig 6.29   Single sideband frequency translation using the filter method

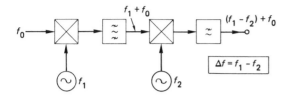

Fig 6.30 Two-stage
frequency translator for
small frequency offsets

This limitation can be overcome by using two stages of filter-based frequency translation as shown in Fig 6.30; however the algorithm complexity becomes rather restrictive. An alternative and much more efficient frequency translation process is the quadrature method described below.

Quadrature frequency translation There are two methods of frequency translation based on quadrature processing: the phasing or Hilbert transform technique, and the Weaver technique. Each have distinct advantages over the filtering method in terms of realizing small frequency shifts and reduced aliasing considerations. A schematic diagram of the Hilbert transform technique is shown in Fig 6.31.

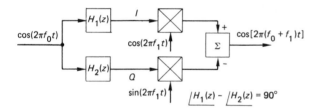

Fig 6.31 Hilbert transform
frequency translator

The circuit performs the mathematical function

$$y(t) = u(t) \cdot \cos \delta\omega t \pm \hat{u}(t) \cdot \sin \delta\omega t$$

where $u(t)$ is the signal to be translated, $\hat{u}(t)$ is the Hilbert transform or quadrature version of the message signal, and $\delta\omega$ the desired frequency shift. Each multiplier generates an upper and lower sideband term (Fig 6.31). However, one of the sidebands cancels out at the final combiner stage, depending on the 'sign' of the combiner. If the two terms are *added*, then the lower sideband is selected and *downwards* frequency translation is achieved. If they are *subtracted*, the message is translated *upwards* in frequency.

Undistorted frequency translation relies on perfect implementation of the above expression. The quadrature versions of the input signal $u(t)$ can be generated by any of the techniques described in section 6.3, with the Hilbert transform FIR filter being the most common approach. In practice, there is usually some error (amplitude or phase) in the quadrature waveform generation process with

Amplitude error (dB)	Phase error (degrees)	Sideband suppression (dB)
3	0	15
1	0	25
0.1	0	45
0.01	0	65
0.001	0	85
0	30	23
0	20	30
0	10	46
0	5	54
0	1	82

Table 6.5 Sideband suppression as a function of amplitude and phase error

the result that the unwanted 'sideband' from the mixing process is not completely suppressed. Table 6.5 gives the theoretical sideband suppression for various values of amplitude and phase error.

The ability of the Hilbert transform technique to realize small frequency shifts is brought about by the elimination of the output filter stage. The requirement still remains, however, for a low-frequency quadrature oscillator. One possible algorithm for this generator is the look-up table method outlined in section 6.2. However, a more desirable implementation is the recursive filter technique. Unfortunately, this realization cannot be easily used for low-frequency oscillators and is thus not directly applicable. To allow the use of recursive oscillator techniques, a double mixing quadrature frequency translator can be implemented as shown in Fig 6.32, where the frequency shift corresponds to the difference frequency between the two local oscillators.

The main drawback of either of the above techniques is that any residual sideband components resulting from imperfect processing lie outside the frequency band occupied by the translated waveform (Fig 6.33) and can cause distortion to neighbouring signal components.

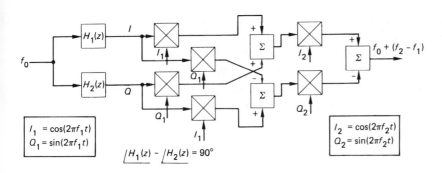

Fig 6.32 Hilbert transform frequency translator for small frequency offsets

$$I_1 = \cos(2\pi f_1 t)$$
$$Q_1 = \sin(2\pi f_1 t)$$

$$\underline{/H_1(z)} - \underline{/H_2(z)} = 90°$$

$$I_2 = \cos(2\pi f_2 t)$$
$$Q_2 = \sin(2\pi f_2 t)$$

$$f_0 + (f_2 - f_1)$$

327

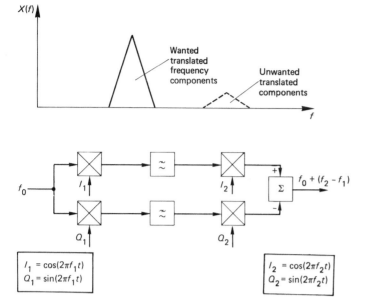

Fig 6.33 An example of adjacent spectral pollution due to poor sideband suppression

$X(f)$

Wanted translated frequency components

Unwanted translated components

f

Fig 6.34 Weaver method of frequency translation

f_0

I_1

$\sim$

I_2

Q_1

$\sim$

Q_2

Σ

$+$

$-$

$f_0 + (f_2 - f_1)$

$I_1 = \cos(2\pi f_1 t)$
$Q_1 = \sin(2\pi f_1 t)$

$I_2 = \cos(2\pi f_2 t)$
$Q_2 = \sin(2\pi f_2 t)$

The problem is solved by using a frequency conversion process known as the Weaver method (cf. section 6.3). This technique also removes the need for wideband quadrature waveform generators. A schematic of the Weaver system is given in Fig 6.34.

The Weaver technique operates by first down-converting the input signal using quadrature local oscillators, with the local oscillator frequency chosen such that the lower sideband folds approximately symmetrically about zero hertz. After filtering to remove the upper sideband from the mixer process, two quadrature lower sideband signals remain. These two signals are then modulated with a further pair of quadrature oscillators in an identical manner to the Hilbert transform method described above. When the mixer outputs are combined, one of the sidebands from each mixer is suppressed (assuming perfect quadrature oscillators), giving the desired frequency translated output waveform. The translation frequency is simply the difference in frequency between the two quadrature oscillator pairs.

The Weaver technique as described has two major attractions compared with the Hilbert transform method. Firstly, no wideband $90°$ phase shift network is required, since the processing is accomplished using quadrature local oscillators which can be very accurately derived (cf. section 6.2). This means that the likelihood of poor sideband suppression due to quadrature imbalance is greatly reduced. Secondly, the Weaver method has the property that if poor sideband suppression does occur (possibly due to a trade-off of processing time and waveform accuracy), the unwanted sideband is

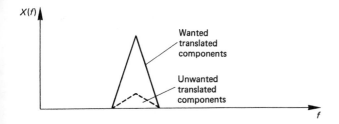

superimposed on the wanted sideband and does not occupy spectrum which may be allocated to another signal (Fig 6.35). In many communications and signal processing applications, this is much more acceptable than the spectrum pollution resulting from the Hilbert transform or phasing method of frequency translation.

Maintaining quadrature In some applications, for example quadrature frequency translation, it may be desirable to retain quadrature versions of the processed signal. This can be simply achieved by performing a cross-coupled mixing process as shown in Fig 6.36. The processing overhead involved is minimal.

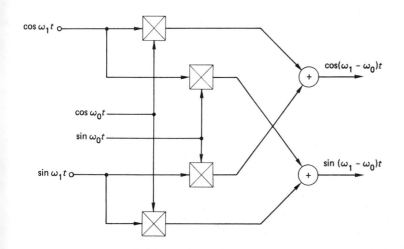

Fig 6.36 Quadrature frequency translation maintaining quadrature output components

Frequency translation by aliasing A further technique for frequency translation is to make use of the aliasing phenomena of sampled systems—more often a hindrance than a help! We know that any signal which has a frequency component greater than half the sampling frequency will be aliased somewhere within the baseband bandwidth of 0 Hz to $f_s/2$ Hz. Therefore, if we wish to down-convert the signal shown in Fig 6.37a, all we need do is to ensure that a harmonic of our sampling rate coincides with the minimum signal frequency component (Fig 6.37b) and the conversion is automatically accomplished. The alias component can subsequently be

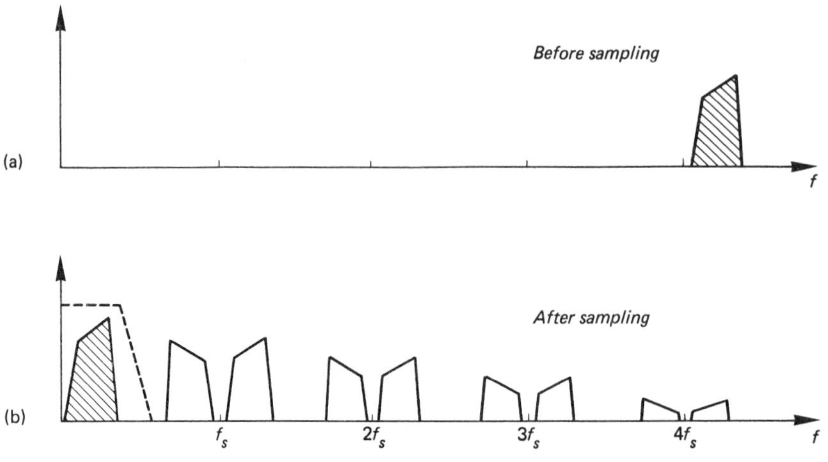

Fig 6.37 An example of frequency downconversion using aliasing

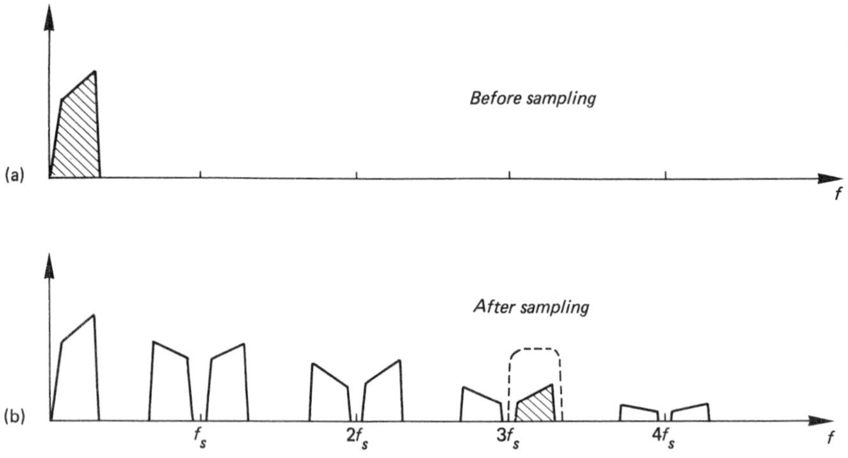

Fig 6.38 An example of frequency upconversion using aliasing

isolated by filtering. (If quadrature down-conversion is required, then two sampling chains can be used, one in quadrature with the other.)

Exactly the same reasoning can be applied for frequency up-conversion (Fig 6.38). A sampling rate is chosen that results in an alias component falling in the desired higher-frequency band, which is selected by a suitable bandpass filter.

Where the sampling rate is fixed by some external or internal factor, interpolation and decimation techniques can be used to obtain the required effective sampling rate for the degree of frequency translation sought, cf. section 4.7.

Care must be exercised when using aliasing for down-conversion, to ensure that only the desired components are aliased to baseband. This usually entails some kind of preselection filter. Also, for

330

down-conversion of very-high-frequency signals, fast sample-and-hold amplifiers are necessary to accurately capture the instantaneous waveform level.

The major disadvantage with the aliasing technique for up-conversion is that the magnitude of the aliased component decreases as the degree of frequency translation attempted is increased.

Frequency inversion A special case of frequency translation is frequency inversion, whereby the complete band from 0 Hz to $f_s/2$ is inverted. The process is in fact trivial for a sampled waveform, and involves mixing by a component at $f_s/2$. This results in upper and lower sidebands, the upper sideband folding exactly on top of the lower sideband which is reversed—hence frequency inversion. Mixing by $f_s/2$ can be achieved by inverting every other waveform sample, i.e. multiplication by $(-1)^n$. An obvious application of frequency inversion is for a basic voice privacy system.

6.7 Signal averaging

Signal averaging is in reality a filtering operation on selected waveform samples. Many of the algorithms for signal averaging are thus covered in Chapter 4 on digital filtering. Three types of signal averaging are commonly employed: linear averaging, peak averaging, and exponential averaging.

Linear averaging Linear averaging simply entails finding the mean of a number of sample values, $x(n)$, as defined by the equation

$$L_{av.} = \frac{1}{N} \sum_{n=1}^{N} x(n)$$

Clearly, this calculation involves ADDition of the N samples, and subsequent division by N. This algorithm is greatly simplified if N is a power of two, i.e. 2, 4, 8, 16..., since the division (normalization) is accomplished by a right shift of the appropriate power of two.

Exponential averaging In contrast to linear averaging, where all samples are summed with equal weighting, exponential averaging uses non-uniform weighting, with greatest weight usually given to the most recent samples. In this way, the averaged output is biased towards the more recent changes in a waveform and can thus 'follow' slowly varying changes in signal shape. As mentioned earlier, averaging is in essence a filtering operation and as such can be implemented using recursive or non-recursive algorithms.

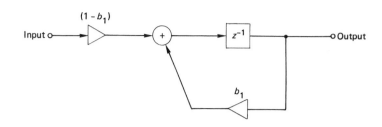

Fig 6.39 Simple exponential averaging circuit

A simple feedback filter structure is shown in Fig 6.39, and is in fact the digital realization of a first-order low-pass filter. The weighting of the accumulated samples, past or present, is determined by the feedback constant b_1. The larger the value of b_1, the greater the emphasis on past history, whereas the smaller b_1, the greater the emphasis on present samples. These techniques were discussed more fully in Chapter 4.

Peak averaging   Strictly speaking, peak averaging is not really an averaging process at all. All that is involved is detecting and storing the peak level of a particular signal component. It is a common feature of spectrum analysers where the peak averaging option results in a display of the peak values of each frequency component occurring over a given period of time. A very simple algorithm can be adopted for this purpose, namely the comparison of present and previous stored samples, with the larger of the two samples being stored.

References

6.1   Peterson W.W. and Weldon E.J., *Error Correcting Codes*, 2nd Edition, MIT Press (1972).
6.2   Gardner F. M., *Phaselock Techniques*, 2nd Edition, Wiley (1979).

7 Hardware/ software device support

7.1 Introduction

The previous six chapters have dealt almost exclusively with signal processing theory and algorithm development, with little reference to hardware and software device support. This final chapter deals with this topic in three sections. It begins with a general discussion on hardware interface circuitry for information transfer to and from the digital environment of the DSP device. The second section deals with the 'higher-level' device support such as software ASSEMBLERS, LINKERS, hardware EMULATORS, etc. The concluding section contains a discussion on likely trends in DSP technology of the future.

7.2 Hardware interface requirements

In all DSP applications, the processor must interface to the outside world, whether it be in the form of A/D or D/A converters, program memory access, digital I/O, external memory addressing, interrupts, timing, etc. A detailed discussion of specific device interconnections and timing peculiarities for every peripheral configuration is clearly not viable in the space available. However, some general comments can be made. When read in conjunction with the appropriate manufacturers data sheets and standard texts on microprocessor interfacing, the information presented should shorten the hardware design phase for DSP considerably.

7.2.1 Digital interface

Processor bus considerations The interfacing of a DSP to peripheral devices almost inevitably involves the common use of data buses, address buses and control signals (Fig 7.1). Many, but not all, processors use a common data bus for external data memory, program memory, digital I/O and multiprocessor communications,

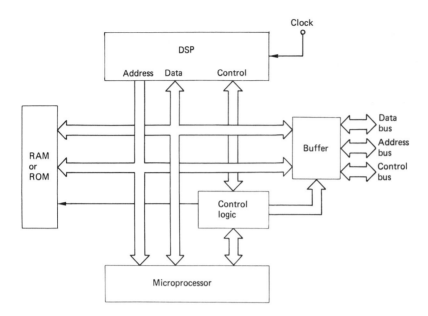

Fig 7.1 Multi-bus DSP
system configuration

and it is imperative that no two devices are driving a common bus at the same time. To satisfy this requirement, the majority of peripheral devices are equipped with an output enable/disable control. When disabled, the outputs of the device are placed in a high impedance mode, which effectively releases the bus for use by another peripheral. The more recent DSP products permit disabling (or tri-stating as it is called) of their own data interface, allowing peripherals to communicate with each other whilst sharing a common bus with the processor. Quite clearly, the time taken by a device to access and use the data bus will determine how effectively other peripherals can operate.

With a single address bus and data bus, the processor can access only one peripheral device at any given time, which obviously restricts the rate of information transfer. This limitation has prompted the development of processors with multiple external address and data buses with a consequent doubling or tripling of the pin count. Unfortunately, this entails a significant increase in cost and complexity of PCB design and manufacture.

External memory The three common types of memory available are Random Access Memory (RAM or DRAM), Programmable Read Only Memory (PROM) and Erasable Programmable Read Only Memory (EPROM), the latter two often referred to simply as ROM. RAM is the only device which can be dynamically written to as well as read, and is thus the most flexible for development purposes and volatile data storage. However, unless a battery backup is available, RAM cannot provide a permanent memory store and ROM must be used. EPROM allows the memory contents to be altered by a process of ultra-violet erasure and re-programming, and is thus considerably more versatile than PROM. Electrically erasable programmable read only memory (E^2PROM) is more versatile still, as the memory can be erased and reprogrammed in situ. (EPROMs and E^2PROMs with access times of only a few nanoseconds are now commercially available at reasonable prices.)

Program memory As with any microprocessor device, the DSP must address and access program memory. A number of DSP manufacturers are marketing products with mask programmable program memory on-chip, which for volume applications is the most logical and efficient means of program memory access. For small-volume applications and research and development design phases, this is at present not a cost-effective or practical solution. Manufacturers are beginning to release DSP parts with on-board fast EPROM-based program memory and even E^2PROM memory, which is a significant step towards promoting low-cost compact applications of DSP technology. Examples are the TMS320C15 and TMS320C17 parts from Texas Instruments and the DSP320EE12 from GI. A few manufacturers are providing a limited amount of on-chip RAM which can be dynamically configured as program memory.

When there is a need for external program memory, the important parameters for consideration are program memory size and access speed. The more recent DSP devices typically have a 16-bit (or greater) bus which can thus address up to 64K words of program memory. This, for most applications, is more than ample, but if extended program memory is required then it is conceivable to switch between successive banks of 64K memory under software control. Many of the more recent processors also allow spare external program memory to be configured as data memory when required.

As the clocking rate of DSP devices increase, so the access time of external memory must be decreased. With single cycle execution times of less than 55 ns for some of the newer processors (e.g. AT&T's HE DSP16), memory access must not exceed about 30 ns for reliable information transfer.

When processor speed is not a limiting factor, memory access times can be relaxed in one of two ways. Firstly the device clock rate can usually be reduced (selecting a lower frequency crystal), thereby slowing down the processor cycle time and relaxing memory access times accordingly. Secondly, with the processor running at full speed, some devices permit a number of 'wait' states to be incorporated into the program memory access cycle, increasing the program memory access time allowed. Reference should be made to the device user manual to exploit this facility. A good example of the potential of the wait state facility is found on some of the more recent TMS320 series. These devices have on-chip RAM configurable as program memory allowing small programs to be stored and executed at full speed entirely on-chip. The initial loading of program memory from an external source can be accomplished using wait states and slow EPROM, overcoming the need for expensive fast ROM and without sacrificing ultimate processing speed.

Unless the processor has a dedicated program memory data bus, the external memory must be disabled after every read/write cycle to allow other peripherals, e.g. data memory or A/Ds, to access the data bus. The processor thus provides an enable signal when a program fetch occurs. Some devices also provide a read/write signal for use with RAM-based program memory.

Data memory Almost all current DSP devices feature on-chip RAM configured as data memory. For many applications, this is sufficient, and access to external data memory unnecessary. When external data memory access is required, the same considerations as outlined for program memory apply. Primarily, the access time must be less than the processor cycle time if full advantage is to be made of the processor speed, but as for program memory, the use of wait states or a reduced clock rate can accommodate slower, cheaper external memory. Because of the similarities of RAM-based external program memory and data memory hardware, the two are interchangeable and many DSP devices can define external memory as either data or program memory under software control.

Digital input/output Digital input/output encompasses the remainder of the peripheral devices, apart from control circuitry. As for program and data memory, one of the major considerations is access time of the peripherals, with A/Ds and D/As exhibiting particularly slow access rates compared with modern processor cycle times. Wait states can be used if supported by the DSP device, with a resulting speed penalty. Alternatively where wait states are not supported or desirable, a buffer stage can be inserted to which data can be input or output at leisure by the peripheral, and accessed at high speed by the processor. Clearly this involves an additional control overhead in most cases.

The more recent processors provide both a serial and parallel digital I/O facility, the serial port having the advantage of being separate from the data bus and thus not constrained by stringent access times and possible bus sharing considerations.

Device addressing Besides the obvious need for addressing of external memory, there is a need for an addressing mechanism to select and enable peripheral digital I/O devices. This is typically achieved by using some of the lower bits of the memory address bus coupled with some decoding logic. For example, the lower 4 bits are used in the TMS320-20, which can thus address up to 16 peripheral devices with the use of a 4 to 16 decoder chip. One further control signal is needed to differentiate between memory addressing and digital I/O addressing. (If only four external devices need to be addressed, then the four processor address lines can be used directly as enable signals without further decoding.)

Wait states The term 'wait states' arises from the capability of many DSP processors to wait for a number of processor cycles while some time-consuming external interface operation is performed. The most obvious application is when writing or reading from peripheral devices which have access times longer than can be accommodated within a single processor cycle period. Wait states are controlled by an external input to the processor which, when active, places the DSP in a waiting mode until information transfer is complete. Deactivating the control signal causes the processor to continue. (Note: in time-critical loops, the additional time lost due to wait states must be taken into account.) The control signal may be derived from the peripheral device being accessed, or from some external logic which introduces a predefined number of wait states every time a particular device is addressed. In many cases only one wait state is required to satisfy the access time constraints of common peripheral devices.

Control signals The typical control interface of a processor consists of a RESET signal, WAIT or HALT signal, READ/WRITE signals to program and data memory, and digital I/O and interrupt lines. The DSP control signals are either active high or active low, the logic state chosen to marry with the control signals of most common peripheral devices. For some external devices it will be necessary to invert the control signals, or employ additional decoding logic. Reference to the appropriate data manual should provide the necessary circuitry and information.

Multiprocessor operation There are inevitably some applications, particularly in the image processing field, where a single DSP device is not sufficiently powerful to perform a given task. The parallel operation of processors is quite feasible, but requires careful planning of both algorithm partitioning and hardware interfacing to achieve good results. One particular consideration is the synchronization of a number of processors to have identical program execution rates, etc. This is facilitated on the more recent products with the provision of a SYNC output for clock control. An alternative approach is to use interrupts from a designated *master* processor to control the functions of a number of *slave* processors. Manufacturers usually provide the necessary information on multiprocessor interfacing for their particular product.

7.2.2 Analog interface

The interfacing of the DSP to an analog environment involves a number of stages as illustrated in Fig 7.2. The purpose of each stage has been discussed in detail in Chapter 2, together with the relevant sampling theory, and is thus not repeated here. This discussion is concerned solely with the interfacing and implementation of the various stages from a hardware design standpoint.

Fig 7.2 Typical analog interface system

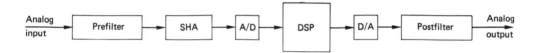

There are two main types of A/D and D/A converter, distinguished by whether the processor digital interface is parallel or serial. The latter type are often called *codecs* and may well incorporate anti-alias/reconstruction filtering and companding, as well as A/D and D/A conversion.

Parallel A/D converters

1 *Resolution and Timing* In general, the faster the conversion time and the greater the number of bits, the more costly the device. 8-bit converters are by far the most common, due to their compatibility with general-purpose microprocessors, and conversion times of about 10 μs are quite feasible (allowing an effective sampling rate of 100 kHz). If greater resolution is required, 10-bit and 12-bit converters are available, although markedly more expensive and slower (typically 10–30 μs). Full-resolution 16-bit converters can be purchased, but at a price! For very fast conversion rates, a range of 'flash' converters or video converters are becoming available, typically with 6 or 8 bits resolution, with operating frequencies of several megahertz.

2 *Data format* Nearly all converters represent the encoded waveform sample using an offset binary format. Conversion to a 2's complement format within the processor is thus essential for most algorithms. A suitable procedure is suggested in section 3.4.

3 *Timing* The access time for converters varies considerably, and is unfortunately in most cases not fast enough to accommodate a single cycle data access from the processor. Wait states or buffering are thus usually required. Similarly, the control signals for conversion start, etc., from the processor are often less than the minimum specified for the converter, and, again, wait stages, latches or even a monostable can be used to extend the control signal duration.

Other parameters such as conversion linearity, supply voltage, temperature stability, cost, power requirement, dc-offset and external/internal clocking should also be taken into account when selecting an A/D device.

Some of the more recent A/D devices incorporate a sample-and-hold amplifier within the same package which is attractive for high-density PCB design.

D/A conversion

1 *Resolution and timing* The cost of D/A converters is considerably less than A/D converters for the same resolution, and the conversion or settling time in the order of a 100–300 ns for 8-bit converters rising to 1–2 μs for 12-bit conversion. 'Flash' or video converters are also available, with operating frequencies of 100 MHz or more.

2 *Data format* For compatibility with the majority of A/D converters, D/A devices accept offset binary formatted data. Conversion from the 2's complement notation of the processor can either be accomplished in software as discussed in section 3.4, or using dedicated hardware. For information on the latter, the reader is referred to the manufacturers data sheet.

3 *Timing* As for the A/D, the write control signal from the DSP is often less than the minimum specified for common D/A devices, as is the time for which valid data is present on the DSP data bus. Wait states, buffering, etc., must thus be incorporated in the hardware design if timing is a problem.

Other factors such as supply voltage, power dissipation and linearity should all be considered.

Sample-and-hold amplifier (SHA) The purpose of the sample-and-hold device is to freeze the analog input waveform whilst the A/D conversion is taking place. This ensures that the binary signal corresponds exactly to the signal amplitude at the sampling instant, i.e. when the sample-and-hold is activated, rather than the average signal level during the conversion period. The characteristics of the SHA are crucial to system accuracy and the reliability of the digital data, especially in applications using 12 or more bits, and where a fast sampling rate is employed.

The important parameters of the sample-and-hold device such as acquisition time, sample accuracy and stability are determined primarily by the storage capacitor in the SHA circuit. To obtain the best performance for the SHA stage, careful reference to the manufacturers specification on capacitor type and value should be made. A poor choice of capacitor will result in a noticeable increase in noise and distortion of a reconstructed analog waveform.

Input anti-aliasing filters Unless aliasing is deliberately intended, e.g. for frequency conversion (Chapter 6), then it is advisable, though not essential, to use an anti-aliasing filter which attenuates any signal components with frequencies in excess of half the sampling frequency. A cost- and space-effective filter can be realized using switched capacitor techniques, and a number of devices are available on the market offering a range of characteristics. The filters require a clock driver which is typically running at 50–100 times the cut-off frequency, and can either be an integral part of the IC or provided externally. In both cases, the filter itself will produce alias components; however these are at such a high frequency, relative to the DSP sampling frequency, that a simple RC filter is usually sufficient to remove the alias terms.

Clearly, passive and active analog filters can be used for input anti-aliasing purposes if appropriate, and circuits can be found in numerous texts on filter design. The choice of filter type depends on the gain, phase and group delay characteristics which are required of the overall system, including both the alias filtering and DSP sections.

There are some applications where a variable sampling rate is required and it is desirable to alter the anti-aliasing filter characteristics accordingly. Under these circumstances, the switch capacitor filters are ideal, since the clocking rate and hence filter cut-off frequency can be made a multiple of the DSP sampling frequency, and varied accordingly.

Reconstruction filter In order to convert the sample values produced by the D/A converter into a time continuous analog format, a reconstruction filter or output anti-aliasing filter is required, ideally having a passband and stopband extending from zero hertz to $f_s/2$ and from $f_s/2$ upwards respectively. As such a filter cannot be realized and an approximation to the ideal response must be accepted. As for the input aliasing filter, the gain, phase and group delay characteristics of the filter are important since they affect the response of the overall system. For most applications, the switched capacitor filter is very suitable as an aliasing filter, particularly if variable sampling rates are anticipated as discussed in the previous section.

If sufficient processing capability is available within the constraints of program length and execution time of a given DSP application, then oversampling on both input and output waveforms can greatly alleviate the constraints on anti-aliasing filter characteristics. Oversampling techniques and associated filtering are discussed in Chapter 4 under the heading of interpolation and decimation.

Serial A/D and D/A converters (codecs)

Codecs are devices which have been developed primarily for the telecommunications market, and incorporate both serial A/D and D/A converters, anti-alias/reconstruction filtering and companding on a single chip (Fig 7.3). As such, they present a very attractive interface device for voice processing applications provided that a serial data I/O capability exists on the processor. If this is not the case, then external serial-to-parallel conversion is necessary, using for example a shift register. Alternatively, the serial-to-parallel conversion can be performed in software. In this case, the data is input on one line of the 16-bit data bus, and the accumulator used to align the data with appropriate left or right shifts. A suitable algorithm is given in Example 7.1.

Fig 7.3 Typical codec block diagram

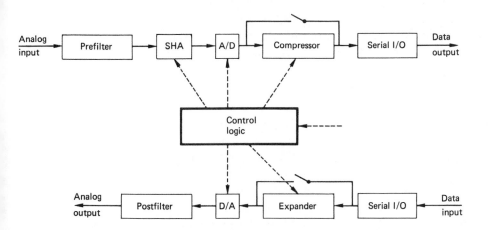

The disadvantage of a codec as a general analog I/O device is that in many cases the sampling rate is fixed, typically at 8 kHz, and it is not possible to disable the filtering and companding. The conversion process is thus non-linear, both in terms of gain and phase response. Secondly, many of the higher specification codecs, e.g. 12-bit and 16-bit devices, incorporate a 'sliding window' technique whereby the conversion accuracy itself may be only 8 bits, but can be maintained anywhere within a 16-bit dynamic range. This is in effect equivalent to a variable gain amplifier prior to the lower-resolution converter. The dynamic range of the device is thus not equivalent to the codec

TMS EXAMPLE No. 7.1

Input from a serial codec, with software conversion to parallel format.

Input to the LSB of the data bus is assumed, with the MSB of the data word received first. Unused lines of the data bus are low during input from the codec.

Now begin main program.

```
START    IN     INPUT,CODEC        :input from codec
         LAC    INPUT,15           :load MSB of low accumulator

         IN     INPUT,CODEC        :input next bit
         ADD    INPUT,14           :align in accumulator

         IN     INPUT,CODEC        :next bit
         ADD    INPUT,13           :align
```

The above process of inputting bits and aligning them in the low accumulator should be repeated for the length of the codec data word.

```
         SACL   INPUT              :save result from accumulator
```

resolution. Further, the harmonic and intermodulation distortion may not be acceptable for non-voice applications.

If a codec with integral μ-law or A-law companding is to be used as the interface for the DSP, then the binary representation of the analog input sample is no longer linearly related to the sample magnitude. Thus, in order to perform linear processing of the sampled analog signal, e.g. filtering, an 'inverse' companding process must be performed within the processor to restore a linear relationship. Similarly, the converse process must be performed on the data being output to the codec to obtain the companded form expected by the output device.

In some of the more recent DSP devices with dedicated serial interface ports (e.g. TMS320-17), the linearization (expansion) of the input data and complementary compression of the output data is performed in dedicated hardware, under software control.

When this facility is not available, the linearizer must be implemented entirely in software, based on either the curve fit or look-up table techniques, as outlined in Chapter 3. This inevitably involves a processing overhead which should be taken into account when choosing the analog I/O devices. Suitable algorithms can be found in Ref 7.1

The codec, it must be stressed, is a special type of serial A/D and D/A converter, and there are many products on the market which provide simply an A/D or D/A function alone, with identical performance to their parallel counterparts. For these devices, there is no companding, and hence none of the overheads associated with codecs.

PCB layout considerations   There are two essential factors to bear in mind when laying out a PCB. Firstly, keep the analog and digital sections physically separate. Secondly, place components associated with particular analog ICs as close as possible to the actual device, thereby minimising stray capacitance effects and pick-up. Decoupling of power supply lines at the DSP itself is essential for low noise and reliable processor operation. The same applies to the analog and interface components. With the multi-bus structure and consequent high pin density of modern DSP ICs, double-sided, if not multi-layer, PCB manufacture is essential.

7.3   General-purpose analog interface board

A number of DSP manufacturers and third-party support companies market general-purpose analog interface boards which include most of the above component blocks and provide a quick means of real time evaluation and testing of DSP algorithms. A suitable interface board for the Texas Instruments processor is detailed below.

Prototype analog I/O board   Figs 7.4 and 7.5 give the schematic and circuit diagrams of a simple analog interface and DSP board for the TMS320-20 processor.[*] The system provides a single analog input channel and three output channels, each with 12 bits resolution. It also includes all the analog and digital hardware required to realize a high-quality DSP evaluation system. The input sampling rate is restricted to about 40 kHz by the conversion time of the A/D used. However the analog output section can accommodate signals with frequencies in excess of 100 MHz. The sampling rate or program execution time is determined solely by the execution time of the algorithm as no external timing mechanism is used. This approach saves on hardware, but requires suitable padding of algorithms to achieve a desired sampling rate. (Provision is made within the TMS320-20 processor for internal timing, but is restricted to a multiple of four times the processor cycle time.) An example program for analog I/O is given in Example 7.2.

---

[*]Further information on the development system can be obtained from Dr. A. Bateman, Dept. Electrical and Electronic Engineering, Queens Building, Bristol University, Bristol, England. Tel. (0272) 303104.

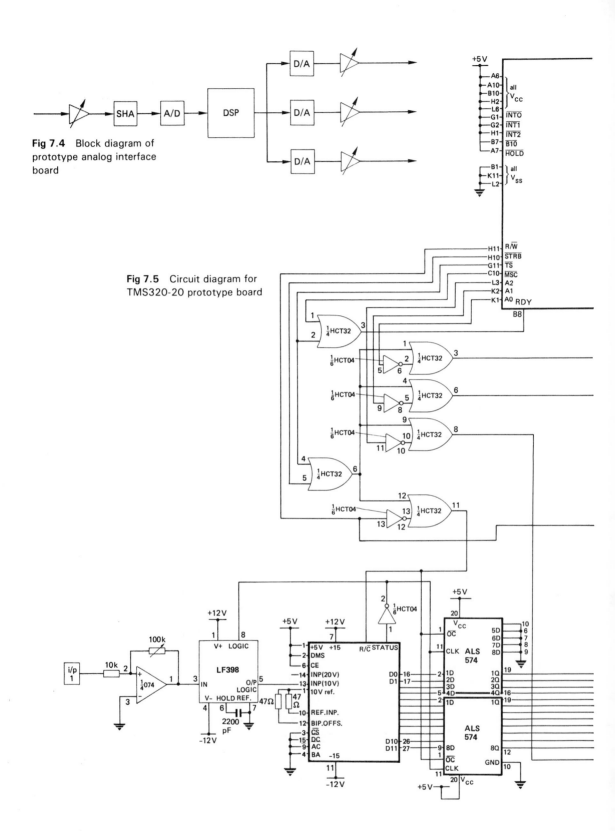

Fig 7.4 Block diagram of prototype analog interface board

Fig 7.5 Circuit diagram for TMS320-20 prototype board

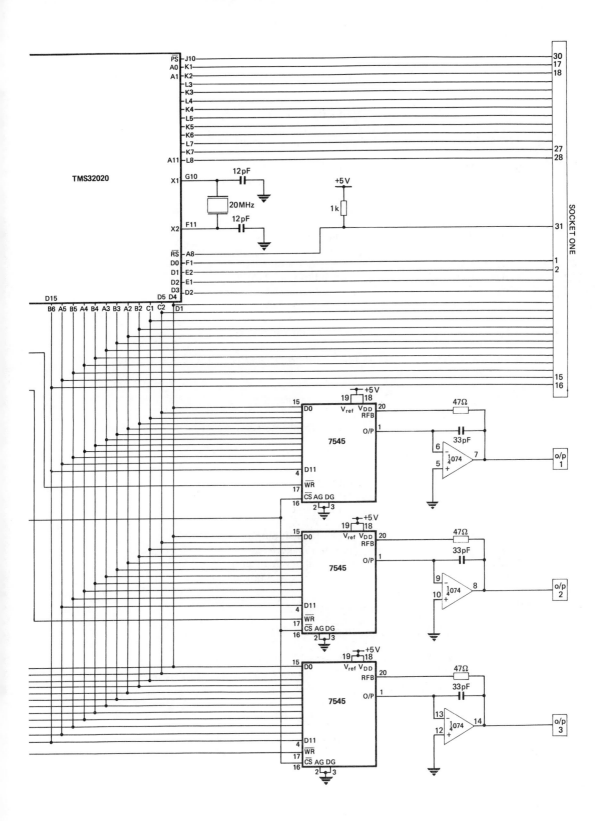

345

TMS EXAMPLE No. 7.2

Analog input and output example for general purpose analog IO board. Sampling frequency is 8 kHz.

```
        IDT     'INOUT'             :give the program a name

        AORG    0                   :set program code start address

        B       INIT                :send resets to a handler
        B       INTRPT              :send interrupts to a handler
```

Declare symbols used to reference data memory addresses, and constants.

```
INPUT   EQU     96                  :to hold input data

ADC     EQU     PA0                 :port address of ADC
DAC1    EQU     PA0                 :port address of DAC1
DAC2    EQU     PA2                 : "      "     " DAC2
DAC3    EQU     PA4                 : "      "     " DAC3

DELAY   EQU     151                 :delay count for 8 kHz
```

Hardware interrupts are redirected to the following interrupt handler – this is just a dummy routine in this example.

```
        AORG    >20                 :place after vectors

INTRPT  RET                         :return from interrupt
```

Resets are redirected to the following initialisation routine.

```
INIT    DINT                        :disable interrupts
        SOVM                        :set overflow mode on
        SSXM                        :set sign extension on
        CNFD                        :block B0 is data memory
        SPM     0                   :no shift on P output
        LDPK    0                   :use data page 0
        LARP    AR1                 :use auxiliary register 1
        CNFP                        :block B0 is program memory
```

Now begin main program.

```
START   IN      INPUT,ADC           :input from ADC

        OUT     INPUT,DAC1          :output without processing
        OUT     INPUT,DAC2          :similarly for DAC2
        OUT     INPUT,DAC3          :and DAC3

        LRLK    AR1,DELAY           :load delay counter
        NOP                         :padding
PAUSE   NOP                         :padding
        NOP
        BANZ    PAUSE               :delay to give 8 kHz sampling

        B       START               :back for next input

        END                         :declare the program end
```

The potential of this general-purpose analog interface board can only be realized if a flexible means of program development can also be provided. Clearly this is not the case if PROM must be blown for every test program. Even with extensive software simulation and debugging, it is desirable to test and evaluate a system in circuit before committing the program to PROM storage. To achieve this objective, either emulators or RAM-based program memory modules must be used. To complement the general-purpose analog board described in this section, a simple cost-effective, RAM based program memory module is described in section 7.5.

7.4 Program development aids

As stated in the introduction to this chapter, the bulk of the material presented in the book is concerned with the construction of algorithms to achieve a desired signal processing function. Once a suitable algorithm or set of algorithms has been found, the remaining task is to develop and test the program. At the very least this involves the use of a video terminal and keypad for text editing, an assembler for converting assembly code to the object code recognized by the DSP, and either a software simulator package or a hardware evaluation facility to establish the validity of the algorithm. It is conceivable, and in fact was the case for the first Texas processor, that no intelligent computing facility is required, with text editor, assembler and evaluation all performed in hardware serviced by a 'dumb' terminal. By far the more elegant and cost-effective solution for those with access to reasonable desktop computing facilities, and the only path available to those wishing to make use of the more recent DSP devices, is to perform the text editing and assembling in software. This assumption of a reasonable computing base, typically IBM PC (or compatible) or VAX, means that software simulation tools can be provided as well as macro libraries, linkers, algorithm libraries, etc., which facilitate the design, development, testing and evaluation of DSP applications without recourse to any hardware facility at all.

The predictable nature of digital signal processing means that software development can, with a very high degree of certainty, predict and evaluate the actual performance of the physical system. At the end of the day however, algorithms and systems must be tested in real time and as part of the final application circuit, which calls for hardware emulation facilities.

In the following sections, a description of the operation and applications of the various program development aids is given to help the engineer decide on the applicability of a particular design tool for immediate and long-term needs.

347

Text editor There are numerous general-purpose word processing packages which can be used for creating and editing source files that are compatible with the various DSP assemblers. Care must be taken however when using certain packages because they generate control characters incompatible with the assembler. Editing source files in non-document mode generally eliminates most suspect control characters. Remember: always document the source file listing with numerous comment statements.

Assembler The task of the assembler is to convert the source code modules, written using assembly language mnemonics, into executable object code modules compatible with the DSP device in question. In addition, a listing file is usually generated which documents the assembler operations on a particular module or group of modules in terms of data program memory locations, assembly errors, etc. Assembler directives and operation differ for each DSP device, but most incorporate macro library facilities and, in conjunction with a software linker, cater for a modular programming approach using relocatable code. Also provided is a comprehensive assembly error diagnostic package.

Linker The operation of the linker is to combine separately generated object code modules and associated files to form a single 'linked' object code module. The advantage of the modular assembly-linker process is that only the relevant sections or modules of a program need be edited and assembled during program development, rather than the complete source file. This facility can represent a considerable saving in time and effort. A number of special assembler directives and mnemonics relating to memory mapping and organization must be used when adopting the modular approach according to the manufacturer's documentation.

Software simulator Once a complete object code file has been generated, the next stage in the development cycle is program evaluation, either by means of hardware emulation or software simulation, or both. One of the most important DSP design aids is a good comprehensive simulation package. Most packages permit the monitoring of data memory, program memory, auxiliary registers, stacks, etc., with user-definable breakpoints, and display modes. Data can be read from, and output to, designated files for further processing or printing and the contents of registers, memory and program counter modified accordingly. One of the most important features of a simulator is the display facility, with graphics or plotting routines an invaluable means of information transfer.

There are two major limitations of software simulation. Firstly, the simulation is non-real time requiring lengthy simulation runs to establish the steady state performance of feedback control systems such as phase locked loops, etc. The second significant limitation is the lack of hardware emulation, primarily the communication to, and control of, peripheral devices. These two limitations are overcome with the use of hardware emulation systems.

Hardware emulation Two approaches to hardware emulation can be adopted, both allowing real time evaluation of DSP algorithms, and both capable of testing the hardware support. The more conventional and costly approach is to use a self-contained emulator system which interfaces to the application circuit via a header lead and can replicate all the functions of the actual processor. The emulator, controlled by a computer terminal, can also perform most of the functions of the software simulator, including single step program execution, memory/register tracing, etc., and also permits internal or external clocking of the device and, most importantly, hardware debugging. Future systems are likely to incorporate some form of logic analysis capability to further enhance the hardware debugging process.

The alternative approach to hardware evaluation, although not as flexible, is to place the DSP device in circuit, and use flexible program memory to permit testing and modification of real time system performance. The advantage of this approach is the low cost compared with full-blown emulation systems, since RAM-based memory and simple computer interface circuitry is all that is required. Such a system is described in the following section which caters for $4K \times 16$-bit program memory emulations, and uses the RS232 serial communication facility for computer interface. The major restriction of the above approach is that, with no information transfer from program memory to computer, there is no provision for memory/register interrogation. In most cases, this particular operation can be performed adequately using software simulation and is thus not a serious constraint.

7.5 Prototype program memory interface board

As outlined above, much of the hardware design and testing phase can be accomplished using a simple low-cost program memory interface board, as shown in Figs 7.6 and 7.7. The main components of the interface circuit are 4K by 16-bit words of buffered fast-access static RAM (easily extended if required) and a Universal Asynchronous Receiver Transmitter (UART) integrated circuit. The RAM emulates

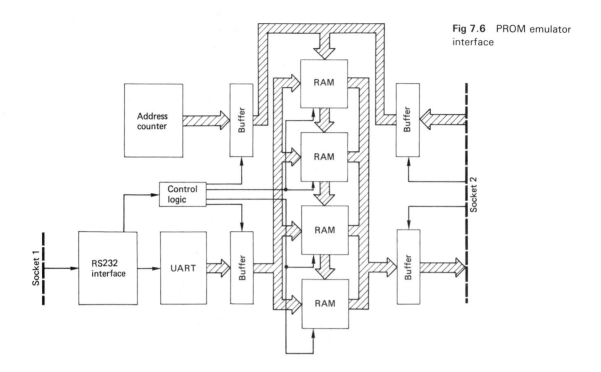

Fig 7.6 PROM emulator interface

PROM ICs to act as program memory for a DSP device through a suitable header connector. The RAM is programmed from the computer through a serial link implemented by the UART and line receiver interface circuits.

The PROM emulation system can be used with any make of DSP device that incorporates a 16-bit-wide program instruction set. The memory access rate is dictated primarily by the specification of RAM used, and is typically 50 ns for the circuit shown.

Programs for the target system DSP which have been converted into object code format by suitable software assembly are downloaded to the interface RAM via the RS232 serial link. During the downloading process, the DSP is held in reset mode and the data and address buffers for the target system disabled. On completion of downloading, the data and address buffers are enabled and the target processor released from the reset state to run the program. A detailed description of the PROM emulator operation is given in the following.*

---

*Further information on the development system can be obtained from Dr. A. Bateman, Dept. Electrical and Electronic Engineering, Queens Building, Bristol University, Bristol, England. Tel (0272) 303104.

System operation The serial link to a computer is implemented using a 6402 UART, a 4702B baud rate generator, a 2.4576 MHz quartz crystal, and a 1489 line receiver. The baud rate generator is hard-wired to run the UART receiver at 9600 baud—only the receiver is utilised to give a one-way link from a computer to the interface. The received data format for the UART is hardwired to 1 START, 8 DATA, NO PARITY, and 1 STOP.

Four signals are taken from the computer serial interface: GND, DATA (out), RTS, and DTR. The DATA, RTS, and DTR signals pass through the line receiver, the latter two acting as control signals for the interface circuit. The DTR line is the main control line, determining whether the computer or the target TMS320 has access to the interface RAM. The RTS line is used to reset the UART receiver buffer register in readiness for reception of another character. Both lines may be controlled from a computer through software.

The downloading process begins by allowing access to the interface RAM data and address buses by the UART and binary counters. This is achieved by asserting a low level (after the line receiver) on the DTR line, which enables the data and address buffers between the RAM and the UART and counters, and disables the data and address buffers between the RAM and the header connector. The low DTR level also resets the binary counters to zero, sets the D-type to its initial state, enables the UART, and can be used to hold a target processor reset. Toggling the RTS line low and then high (after the line receiver) clears the DR (data received) signal of the UART to a low level. When this state has been achieved, the interface is ready for data to be transmitted from the computer.

After successful reception of a character, the UART receiver register holds the 8-bit character in a parallel form, and the DR signal is set at a high level. Before the next character is transmitted, the RTS line has to be toggled low then high to clear the DR signal. The DR signal is then input to the D-type flip-flop, and the outputs are connected to the enable signals of the data buffers between the UART and the RAM, and the write enable signals of the RAM. The outputs follow the input but at half the frequency, and enable the associated data buffers and RAM accordingly, depending on the levels of Q and $\bar{Q}$. In this way, alternate 8-bit bytes are received through the serial link and transferred to either the upper or lower 8 bits of the 16-bit-wide interface RAM. The Q output of the D-type is also used as the input to the binary counters, causing the RAM address to be incremented after every second byte has been transferred.

On completion of the downloading process, assertion of a high level on the DTR line disables the data and address buffers between

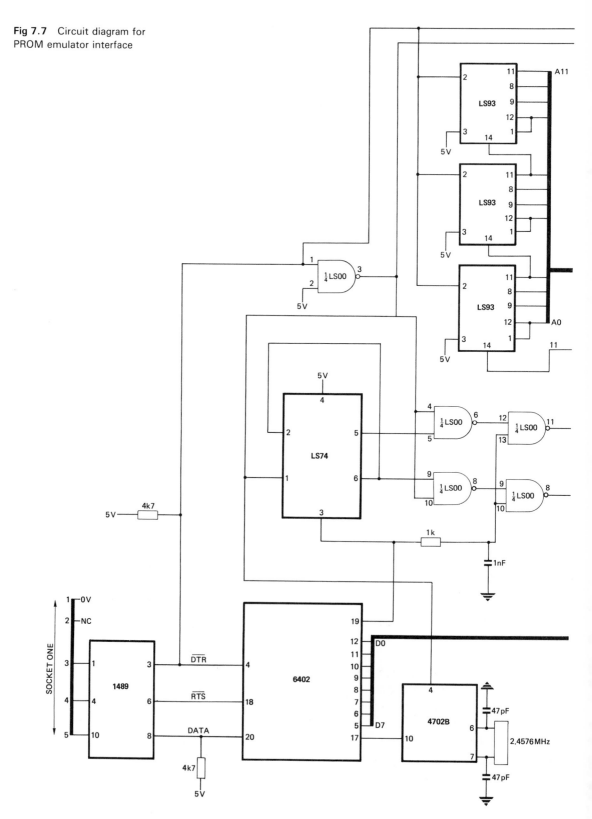

Fig 7.7 Circuit diagram for PROM emulator interface

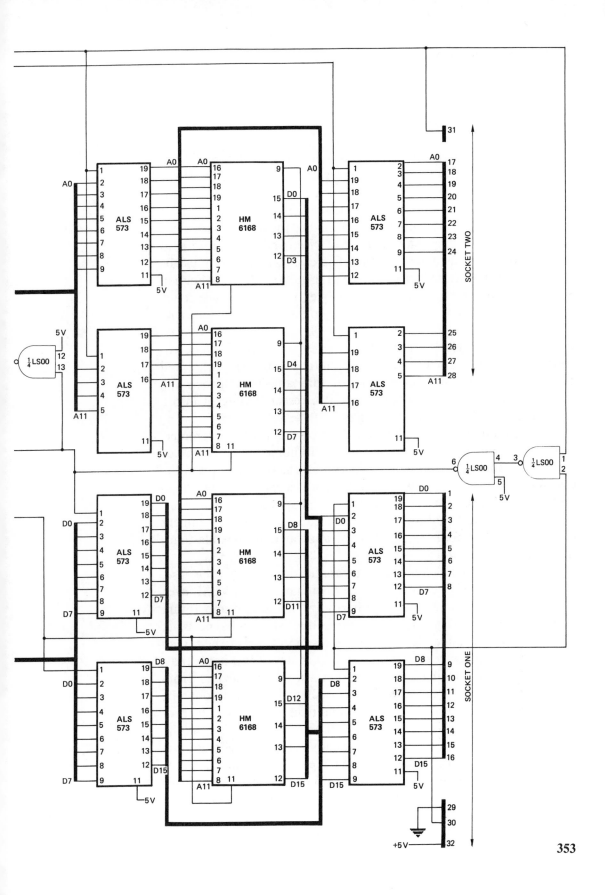

353

the UART and binary counters and the RAM, and enables those between the RAM and the header connector. The target processor would also be released from its reset state.

7.6 DSP—the future?

Prediction of trends in DSP is a risky business, and a path trod warily, but necessarily by manufacturers and users alike. From the manufacturers' standpoint, predicting the correct trends, or attempting to influence them, is essential in view of the massive investment of time and money that must be committed to the development of a new highly-complex specialised DSP device, and furthermore to provide adequate customer support. The commercial user must be able to make an educated guess at the trends in DSP, in order to plan development programs, devise cost strategies, set performance goals, etc., which will ensure the maximum exploitation of the processing power becoming available and thereby obtaining a market edge. From the research standpoint, progress is often hindered, consciously or unconsciously, by the limitations in computing and processing power. An awareness of digital signal processing capability can turn a seemingly impossible research concept into a realistic goal and a commercial success.

There is no doubt that digital signal processing is now a commercial reality and becoming a standard 'workhorse' in all areas of communications, control, image processing, radar, etc. As was the case for the general-purpose microprocessor, second-sourcing, third-party support, and industry standards are all appearing on the DSP scene, and, despite requiring more advanced algorithm development techniques, will become as familiar as, and far more flexible than, the standard microprocessor.

It is quite conceivable, and in some applications already the case, that DSP devices will be more cost effective than custom VLSI, even for large-volume production. The flexibility of modern and future DSP chips means that the applications market will be massive, allowing the high design, development and support costs to be spread very thinly and a low-cost part provided. It is almost inevitable that, after the initial flooding of the market with DSP devices from numerous manufacturers, the market will stabilize and only be capable of supporting two or three dominant suppliers each providing a 'standard' low-cost processor. Of course there will always be a market for specialist devices and already a number of manufacturers are developing products in this area. For example, Texas Instruments are producing a series of specialised versions of their DSP and support chips dedicated to the voice communications and modem markets.

There is clearly a move towards floating-point arithmetic and multibus architectures for future DSPs, although current development and manufacturing costs are high. Power consumption is a serious problem for any complex processor and, despite the CMOS versions of DSPs appearing on the market, there is still a considerable gap between the DSP power requirement and that of a dedicated VLSI approach, the extra power providing the flexibility of a programmable processor. To become a viable alternative to VLSI in a low-power applications environment, sub-micron technology is likely to be essential, and selected power-down facilities on chip utilised. Even so, to maintain a mass market, flexibility must be maintained within the processor design. Redundancy, and hence power loss, is thus inevitable for many applications.

The provision of an analog I/O interface on-chip is certainly a desirable feature for many potential DSP applications, and the absence of such a DSP device is almost certainly hindering the application of DSP in some areas. The problem with integrating A/D and D/A converters with the DSP is, firstly, the large silicon area and power consumption involved and, secondly, the manufacturers commitment to a certain resolution, accuracy and conversion speed for the analog I/O facility, limiting the market in which the device is applicable. This latter factor is paramount in dictating the cost effectiveness of the part.

Device support—the future On the question of device support, there is a rapidly increasing choice of third-party software and hardware available for current DSP devices, ranging from software assemblers, linkers, and simulators for a wide variety of host computer types, through to specialised hardware development tools, plug-in PC boards, etc., as well as numerous algorithm design packages for digital filters, FFTs, speech processing, image processing and so on. A number of companies are supplying high-level language compilers for certain DSPs, using Pascal, Fortran and C—a facility which seems certain to become more widely supported. The drawback of high-level language programming of current DSPs is that the speed and memory capability is not really sufficient to accommodate the inevitable redundancy of a suitably flexible compiler. In the future, when the processor power of DSPs is significantly improved, then such redundancy can most likely be tolerated.

An alternative approach to ease program development is the likely advent of software libraries on-chip. It will, for example, be possible to use a single-device-oriented command which implements a 64-point FFT. Other examples may be sine/cosine generators, basic filter structures, and so on, all permanently resident on-chip.

Whatever the future holds, digital signal processing encompassing the theory, design, development and application, is going to form a major part of modern electronic and electrical engineering. It is hoped that this text has covered most of the basic groundwork involved in the art of DSP, and it is left to the ingenuity of the today's engineer to pursue the DSP applications of the future.

References

7.1   *Digital Signal Processing Applications*, Texas Instruments (1986).

Appendix 1

Summary of First- and Second-order Filter Design

This appendix contains a summary of the coefficients required to realize first- and second-order digital filters. The designs are based on the Impulse Response Invariant and Bilinear Transform methods outlined in Chapter 4. All filters have a maximum gain of unity and are based on the general first- and second-order filter architectures shown in Fig A1.1 and Fig A1.7. Pre-scaling, K1, and post-scaling, K2, are introduced to eliminate overflow when working with fixed-point processors.

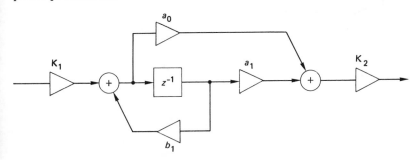

Fig A1.1 Universal first-order filter section

FIRST-ORDER FILTERS

$$H_d(z) = \frac{1 + a_1 z^{-1}}{1 - b_1 z^{-1}}$$

$$|H_d(z)|^2 = \frac{[1 + a_1 \cos(x)]^2 + [a_1 \sin(x)]^2}{[1 - b_1 \cos(x)]^2 + [b_1 \sin(x)]^2}$$

$$\underline{/H_d(z)} = \tan^{-1}\left[\frac{-a_1 \sin(x)}{1 + a_1 \cos(x)}\right] - \tan^{-1}\left[\frac{b_1 \sin(x)}{1 - b_1 \cos(x)}\right]$$

$$(x = 2\pi f T_s)$$

RC low-pass filter: *Impulse Invariant*

$\quad$ K1 $= 1 - b_1$

$\quad$ K2 $= 1$

$\quad$ $a_0 = 1$

$\quad$ $a_1 = 0$

$\quad$ $b_1 = \exp(-T_s/RC)$

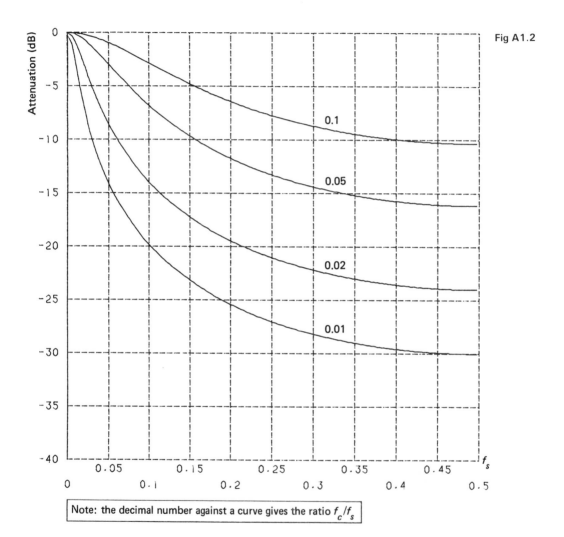

Fig A1.2

Note: the decimal number against a curve gives the ratio f_c/f_s

RC low-pass filter: *Bilinear Transform*

$$K1 = (1 - b_1)/2$$
$$K2 = 1$$
$$a_0 = 1$$
$$a_1 = 1$$
$$b_1 = (1 - T_s/2RC)/(1 + T_s/2RC)$$

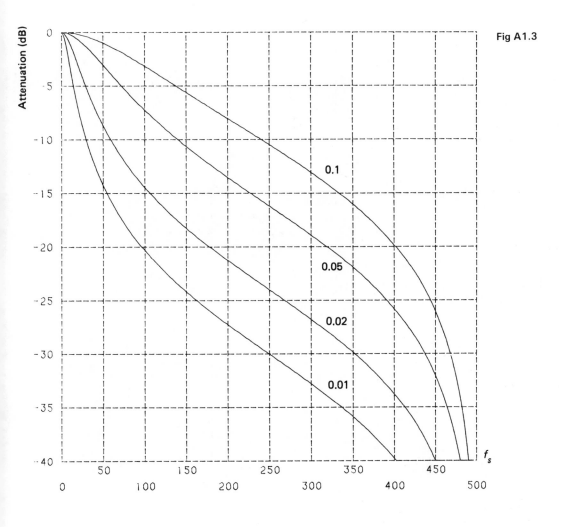

Fig A1.3

High-pass filter: *Impulse Invariant*

$K1 = 1 + b_1$
$K2 = 1$
$a_0 = 1$
$a_1 = 0$
$b_1 = -\exp(-T_s/RC)$

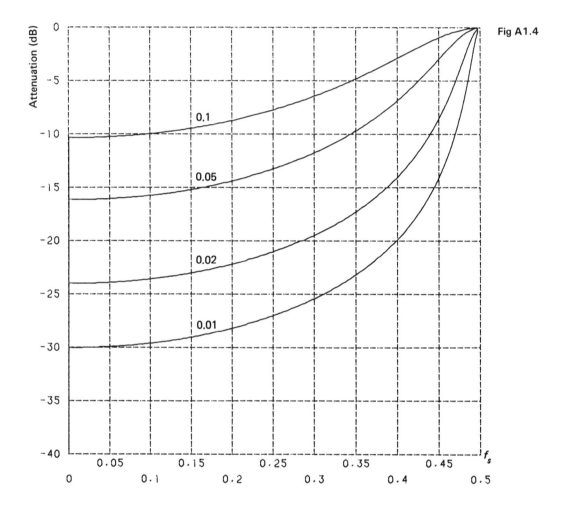

Fig A1.4

CR high-pass filter (AC-coupling): $(1 - LP)$

$K1 = 1 - b_1$
$K2 = (1 + b_1)/2(1 - b_1)$
$a_0 = 1$
$a_1 = -1$
$b_1 = \exp(-T_s/RC)$

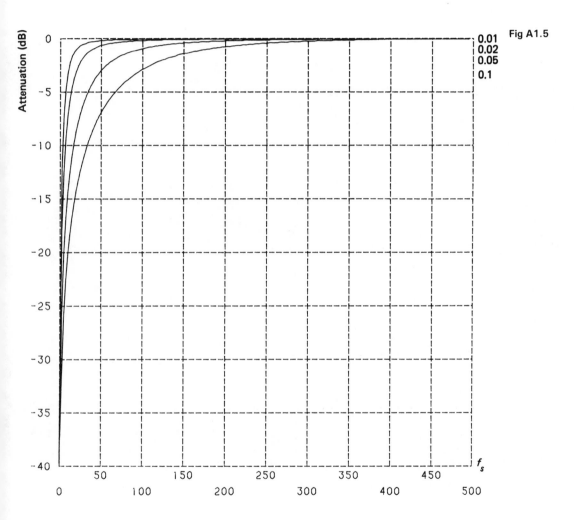

Fig A1.5

CR high-pass filter (AC-coupling): *Bilinear Transform*

$K1 = (1 - b_1)/2$

$K2 = (1 + b_1)/(1 - b_1)$

$a_0 = 1$

$a_1 = -1$

$b_1 = (1 - T_s/2RC)/(1 + T_s/2RC)$

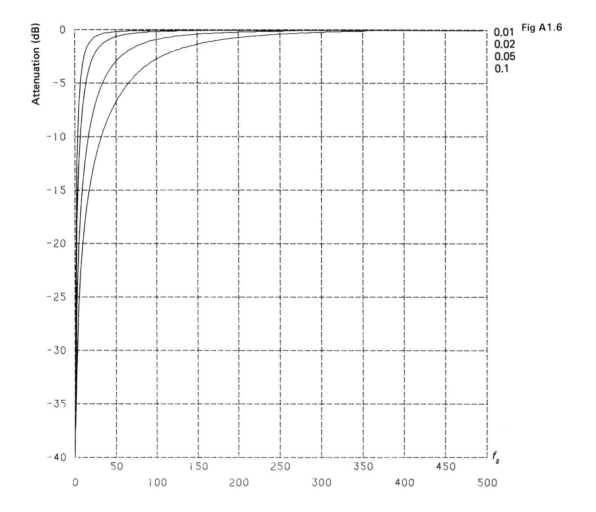

Fig A1.6

SECOND-ORDER FILTERS

$$H_d(z) = \frac{a_0 + a_1 z^{-1} + a_2 z^{-2}}{1 - b_1 z^{-1} - b_2 z^{-2}}$$

$$|H_d(z)|^2 = \frac{[1 + a_1 \cos(x) + a_2 \cos(2x)]^2 + [a_1 \sin(x) + a_2 \sin(2x)]^2}{[1 - b_1 \cos(x) - b_2 \cos(2x)]^2 + [b_1 \sin(x) + b_2 \sin(2x)]^2}$$

$$\underline{/H_d(z)} = \tan^{-1}\left[\frac{-a_1 \sin(x) - a_2 \sin(2x)}{1 + a_1 \cos(x) + a_2 \cos(2x)}\right] - \tan^{-1}\left[\frac{b_1 \sin(x) + b_2 \sin(2x)}{1 - b_1 \cos(x) - b_2 \cos(2x)}\right]$$

$$(x = 2\pi f T_s)$$

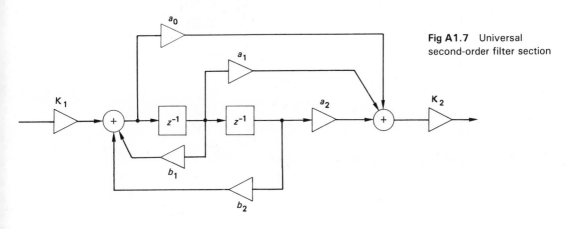

Fig A1.7 Universal second-order filter section

363

Bandpass filter (resonator): *Impulse Invariant*

$K1 = (1 + b_2)/2$

$K2 = 1$

$a_0 = 1$

$a_1 = 0$

$a_2 = -1$

$b_1 = 2 \exp(-\omega_r T_s/2Q) \cdot \cos(\omega_d T_s)$

$b_2 = -\exp(-\omega_r T_s/Q)$

ω_r = angular resonant frequency, $Q = Q$-factor

$\omega_d = \omega_r\sqrt{(1 - 1/4Q^2)} \approx \omega_r$ for $Q > 5$

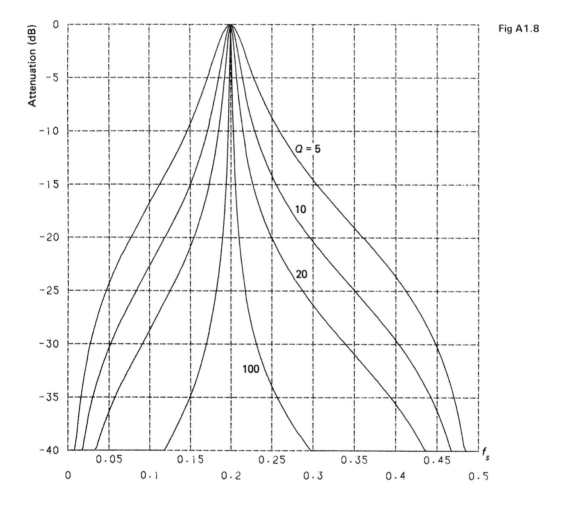

Fig A1.8

Bandpass filter (resonator): *Bilinear Transform*

$K1 = (1 + b_2)/2$

$K2 = 1$

$a_0 = 1$

$a_1 = 0$

$a_2 = -1$

$b_1 = -2(1 - \alpha^2 y)/(1 + \alpha x + \alpha^2 y)$

$b_2 = -(1 - \alpha x + \alpha^2 y)/(1 + \alpha x + \alpha^2 y)$

$\alpha = \omega_r \cot(\omega_r T_s/2) \qquad y = 1/\omega_r^2 \qquad x = 1/\omega_r Q \qquad \omega_r = 2\pi f_r$

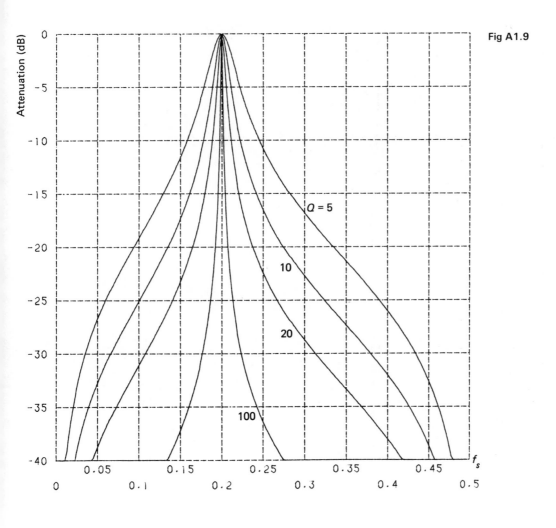

Fig A1.9

Bandstop filter (notch): *Impulse Invariant*

$K1 = (1 + b_2)/2$

$K2 = (1 - b_2)/(1 + b_2)$

$a_0 = 1$

$a_1 = -2b_1/(1 - b_2)$

$a_2 = 1$

$b_1 = 2 \exp(-\omega_r T_s/2Q) \cdot \cos(\omega_d T_s)$

$b_2 = -\exp(-\omega_r T_s/Q)$

$\omega_d = \omega_r \sqrt{(1 - 1/4Q^2)} \approx \omega_r$ for $Q > 5$

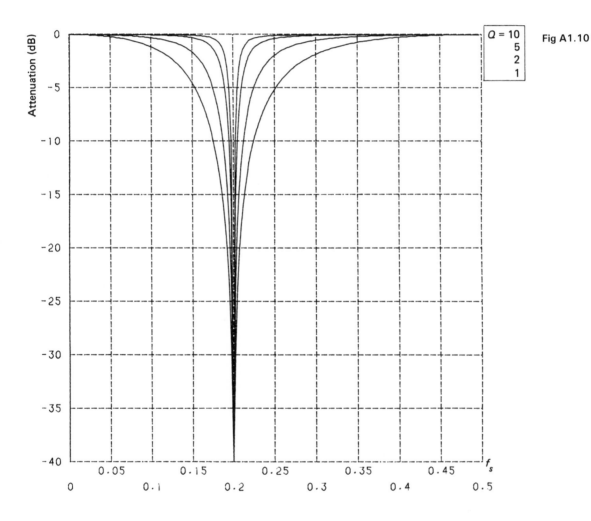

Fig A1.10

Bandstop filter (notch): *Bilinear Transform*

$K1 = (1 + b_2)/2$

$K2 = (1 - b_2)/(1 + b_2)$

$a_0 = 1$

$a_1 = -2b_1/(1 - b_2)$

$a_2 = 1$

$b_1 = -2(1 - \alpha^2 y)/(1 + \alpha x + \alpha^2 y)$

$b_2 = -(1 - \alpha x + \alpha^2 y)/(1 + \alpha x + \alpha^2 y)$

$\alpha = \omega_r \cot(\omega_r T_s/2) \qquad y = 1/\omega_r^2 \qquad x = 1/\omega_r Q$

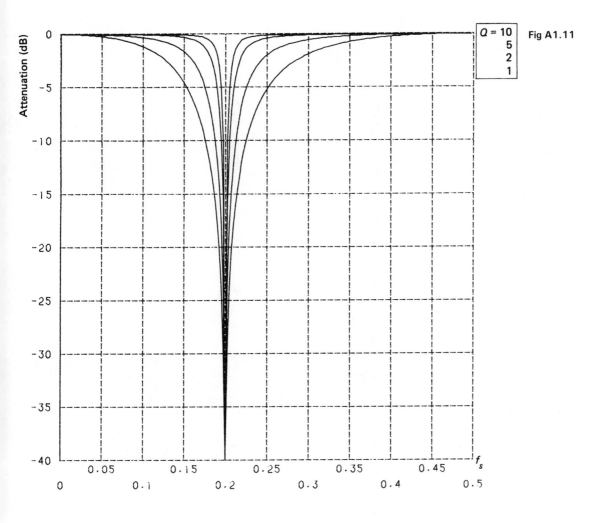

Fig A1.11

367

Allpass filter section (phase equalization): *Impulse Invariant*

$K1 = 1$

$K2 = 1$

$a_0 = -b_2$

$a_1 = -b_1$

$a_2 = 1$

$b_1 = 2 \exp(-\omega_r T_s/2Q) \cdot \cos(\omega_d T_s)$

$b_2 = -\exp(-\omega_r T_s/Q)$

$\omega_d = \omega_r \sqrt{(1 - 1/4Q^2)} \approx \omega_r$ for $Q > 5$

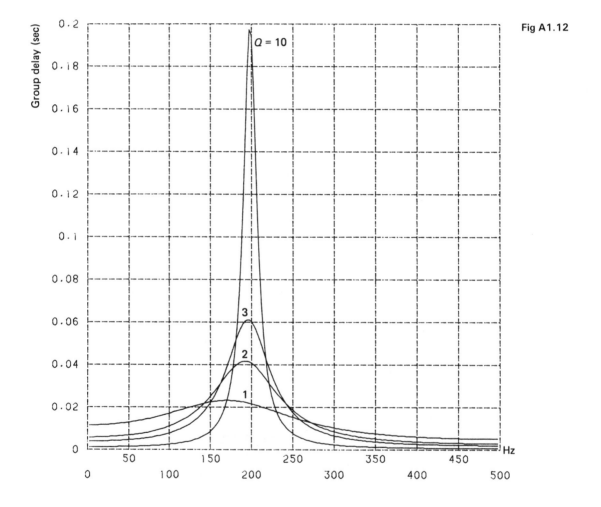

Fig A1.12

Allpass filter section (phase equalization): *Bilinear Transform*

$K1 = 1$

$K2 = 1$

$a_0 = -b_2$

$a_1 = -b_1$

$a_2 = 1$

$b_1 = -2(1 - \alpha^2 y)/(1 + \alpha x + \alpha^2 y)$

$b_2 = -(1 - \alpha x + \alpha^2 y)/(1 + \alpha x + \alpha^2 y)$

$\alpha = \omega_r \cot(\omega_r T_s/2)$ $\qquad y = 1/\omega_r^2 \qquad x = 1/\omega_r Q$

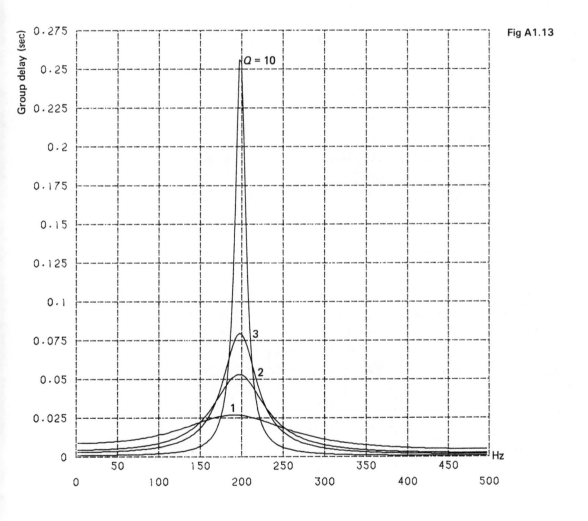

Fig A1.13

Appendix 2

Realization of a 512-point DIT FFT using DSP

The 512-point FFT described in this appendix is an extension of the theory and algorithms derived in Chapter 5. The implementation is tailored towards the TMS320-20 processor, but can be easily modified to suit a number of the more recent DSP products.

For the purposes of this example, it is assumed that the sample values to be processed are held in external data memory and the twiddle factors are defined in program memory.

The first step is to re-order the data samples in external memory in preparation for the DIT FFT algorithm. This can be achieved with the macro given in Example 5.7. If the data commenced at location 1024, the reordering code would be

```
BITREV   2, 512, 1024
BITREV   4, 256, 1024
           ⋮
BITREV   510, 1020, 1024
```

The next step is to read the first block of data into on-chip memory (locations 512 to 1023), carry out the transform, and load the results back into external memory. Assuming a subroutine FFT256 is available to carry out the FFT on the data in locations 512 to 1023, this stage would be coded as:

```
LARP   AR1
LRLK   AR1, 512      Destination in on-chip
RPTK   255
BLKD   1024, * +     Source in off-chip
RPTK
BLKD   1280, * +
CALL   FFT256        Carry out FFT
LRLK   AR1, 1024     Destination in off-chip
RPTK   255
BLKD   512, * +      Source in on-chip
RPTK   255
BLKD   768, * +
```

The second block in the FFT would then be processed in the same way; the only difference in the code would be the replacement of 1024 by 1536 and of 1280 by 1792.

In the final stage, 256 butterflies have to be performed. If speed were critical, the butterflies would also be implemented in two blocks, with the associated data loaded into and out of on-chip memory. Assuming that it is sufficient to simply load the twiddle factors into on-chip memory, the next stage would be:

```
LARP   AR0
LRLK   AR0, 512      Start of twiddle table
RPTK   255
BLKP   W000, * +     Prog MEM start of TT
```

Note that the twiddle factors are assumed to be stored as sine and cosine of the magnitude of the twiddle angle in successive locations of program memory, starting at W000, i.e.

```
W000   DATA   32767, 0
W001   DATA   32765, 402
          ⋮
W255   DATA   − 32765, 402
```

In the final stage, the DIT butterflies are carried out. The twiddle factors are used in order, starting at location 512, with AR1 as the pointer. Similarly, the Q input to the butterflies is used in order, starting at $1024 + 512$, with AR2 as the pointer. AR0 contains the offset between P and Q (the wingspan). It is assumed that a radix-2 DIT butterfly subroutine BFLY is available which expects to find the twiddle and Q pointers in AR1 and AR2 respectively, and the offset in AR0. This routine also expects AR2 to be pointed to on entry, and automatically performs the incrementation of AR1 and AR2. The code is:

```
       LRLK   AR0, 256      AR0 has wingspan
       LRLK   AR1, 512      Start of twiddle table
       LRLK   AR2, 1536     First Q value
       LRLK   AR3, 255      Butterflies to be done
       LARP   AR2
LOOP   BFLY
       LARP   AR3
       BANZ   LOOP, * − , AR2
```

Appendix 3

Derivation of Continuous Signal Correlation and Spectral Density Functions from Samples

Tables are given below describing how the results of discrete operations such as linear convolution, circular convolution and periodic and aperiodic correlation as defined in Chapter 5 (section 5.6) can be used to deduce the correlation and spectral density functions of continuous signals from their ideally sampled versions.

It is assumed that the sampling rate is high enough to satisfy the Nyquist criterion. The relationships for the three cases of interest are given: the input signal being a one-off energy signal, the input signal being periodic, and the input being a random power signal.

In each case there are two routes to the solution, one using algorithms based on the samples themselves and the second based on the DFT of the samples.

Table A3.1 gives the technique for deducing continuous convolutions; Table A3.2 continuous correlation; and Table A3.3 the continuous cross-spectrum density functions.

Table A3.1 Convolution of continuous functions using DSP
(Note: $y(t) = x_1(t) \star x_2(t)$. The function $x_2(t)$ must be finite in length.)

Nature of $x_1(t)$ and $x_2(t)$	Input processing	Direct domain	Transform domain	Output processing
Energy signal (*finite length*)	Sample at adequate rate over length.* (*N* samples)	Discrete linear convolution.	Pad each sequence with zeros to length $2N$. Find DFT of each. Multiply DFTs term by term. Find IDFT = y_n.	Output sequence multiplied by T_s are samples of required $y(t)$. ($2N$ samples)
Power signal, (*periodic*)	Sample at adequate rate over period.*† (*N* samples)	Discrete circular convolution.	Find DFT of each sequence. Multiply DFTs term by term. Find IDFT = y_n.	Output sequence multiplied by T_s are samples of required $y(t)$ over one period. (*N* samples)
Power signal, (*random*)	Sample at adequate rate over block length.† (*N* samples)	Discrete linear convolution of zero padded sequences. (Padded sequence of length $2N$.)	Find DFT of each zero padded sequence. Multiply DFTs term by term. Find IDFT = y_n.	Block processed output sequence (back half of previous block added to front half of current). When multiplied by T_s gives required $y(t)$. ($2N$ samples)

*Either $x_1(t)$ or $x_2(t)$ may dictate the necessary sampling rate.
†The length of $x_2(t)$ must be less than the period or block length.

Table A3.2 Correlation of continuous signals using DSP

Nature of $x_1(t)$ and $x_2(t)$	Input processing	Direct domain	Transform domain	Output processing
Energy signal (*finite length*)	As for convolution. (*N* samples)	Aperiodic correlation.	Pad each sequence with zeros to length 2*N*. Find DFT of each. Multiply first with conjugate of second, term by term. Find IDFT = $\phi(n)$†.	Output sequence multiplied by T_s are samples of required $\phi_{x_1 x_2}(t)$ (2*N* samples)
Power signal, (*periodic*)	As for convolution. (*N* samples)	Periodic correlation.	Find DFT of each sequence. Multiply first with conjugate of second, term by term. Find IDFT, scale by $1/N$ to give $R(n)$†.	Output sequence multiplied by $1/N$ are samples of required $R_{x_1 x_2}(t)$ over one period. (*N* samples)
Power signal, (*random*)	As for convolution. (*N* samples)	Find aperiodic correlation and scale by $1/N$.	As for energy signals, but post scale by $1/N$†.	Block processed output sequence, which when multiplied by $1/N$ gives samples of $\hat{R}_{x_1 x_2}(t)$ (2*N* samples)

†It will be necessary to 'unpack' the IDFT to yield negative time samples.

Table A3.3 Cross-spectral density functions of continuous signals using DSP

Nature of $x_1(t)$ and $x_2(t)$	Input processing	Direct domain	Transform domain	Output processing
Energy signal (*finite length*)	As for convolution. (*N* samples)	Pad each sequence with zeros to length 2*N*. Find DFT of each. Multiply first DFT with conjugate of second, term by term.	Find DFT of discrete aperiodic correlation of two sequences†.	Output sequence multiplied by T_s^2 are samples of the required $G_{x_1x_2}(f)$
Power signal, (*periodic*) (Both must have same period)	As for convolution. (*N* samples)	Find DFT of each sequence. Multiply first DFT with conjugate of second, term by term. Scale result by 1/*N*.	Find DFT of discrete periodic correlation of two sequences†.	Output sequence multiplied by 1/*N* are impulse weights of the required discrete power spectral density function $S_{x_1x_2}(f)$
Power signal, (*random*)	As for convolution. (*N* samples)	As for energy signal, but post-multiply by 1/*N*.	Find DFT of aperiodic discrete correlation of two sequences and post-multiply by 1/*N* (windowed if necessary)†.	Output sequence multiplied by T_s are samples of the estimate of the cross-power spectral density function $\hat{S}_{x_1x_2}(f)$

†The negative time samples will need to be 'packed' into the upper half of the DFT input.

Appendix 4

Derivation of LPC Coefficients from the Autocorrelation Function Estimate

The feedforward coefficients of the all-zero whitening filter for a random signal with autocorrelation function R_n can be found by solving the following matrix equation:

$$
\begin{bmatrix}
R_0 & R_1 & R_2 & \ldots & R_{p-1} \\
R_1 & R_0 & R_1 & \ldots & R_{p-2} \\
R_2 & R_1 & R_0 & \ldots & R_{p-3} \\
\vdots & \vdots & \vdots & & \vdots \\
R_{p-1} & R_{p-2} & R_{p-3} & \ldots & R_0
\end{bmatrix}
\begin{bmatrix}
a_1 \\ a_2 \\ a_3 \\ \vdots \\ a_p
\end{bmatrix}
= -
\begin{bmatrix}
R_1 \\ R_2 \\ R_3 \\ \vdots \\ R_p
\end{bmatrix}
$$

In DSP implementations it is better to use an iterative procedure. Table A4.1 gives a Pascal procedure which converts from autocorrelation function estimate samples to reflection coefficients K_i. The filter can then be directly implemented using a lattice structure. If it is desired to know the value of the corresponding feedforward coefficients the program in Table A4.2 can be used.

```
procedure refcoeff  (r : arr20; (*real array of acf estimate*)
                     p : integer
                     var k : arr20; (*array of reflection coeffs.*)

var
    en, ep : arr20
    ch, h, 1, cm, m : integer;
    flag1; flag2 : boolean;

begin
    for l := 1 to p do

        begin
            en[l] := r[l-1];
            ep[l] := r[l];
        end
    ch := p;
    h := -1;
    repeat

        h := h+1;
        k[h+1] := -ep[h+1]/en[1];
        en[1] := en[1] + k[h+1]*ep[h+1];
        ch := ch-1;
        if ch = 0 then flag1 := true;
        if not flag1 then

            begin
                ep[p] := ep[p] + k[h+1]*en[p-h];
                if ch = 1 then flage2 := true;
                if not flag2 then

                    begin
                        cm := ch-1;
                        m := h+1;
                        repeat

                            m := m+1;
                            ex := ep[m] + k[h+1]*en[m-h];
                            en[m-h] := en[m-h] + k[h+1]*ep[m];
                            ep[m] := ex;
                            cm := cm-1;
                        until cm = 0;
                    end;
            end;
    until flag1;
end; (*refcoeff*)
```

Table A4.1

```
procedure feedfwd  (k : arr20; (*real array of refl. coeefs.*)
                    p : integer;
                    a : arr20; (*real array of feed fwd. taps*)

var
   ci, i, cj, j : interger;
   newa : arr20;

begin
   a[1] := k[1];
   ci := p;
   i := 1;
   repeat

      i := i+1;
      a[i] := k[i];
      cj := i-1;
      j := 0;
      repeat

         j := j+1;
         newa[j] := a[j] + k[i]*a[i-j];
         cj := cj-1;
      until cj = 0;

      for j := 1 to i-1 do
         a[j] := newa[j];
         ci := ci-1;
      until ci =0;
end; (*feedfwd*)
```

Table A4.2

378

Appendix 5

The Autocorrelation Function

The *autocorrelation function* $\phi_{xx}(t)$ is defined as

$$\phi_{xx}(t) = \int_{-\infty}^{\infty} x(t + \tau) \cdot x(\tau)\, d\tau$$

$$= \int_{-\infty}^{\infty} x(\tau) \cdot x(t - \tau)\, d\tau$$

The autocorrelation function is an even function of time with peak value at $t = 0$.

$\phi(0)$ is the waveform energy.

An alternative method of deriving the energy spectral density function is to take the Fourier transform of the autocorrelation function, thus

$$G(f) = \mathcal{F}[\phi(t)]$$

The above results can be extended to encompass periodic signals by scaling the time domain integrals by a factor $1/T_p$, and evaluating over one period to yield the waveform power. To distinguish the power autocorrelation from the energy autocorrelation $\phi(t)$, the symbol $R(t)$ is used. We have

$$R_{xx}(t) = \frac{1}{T_p} \int_{-T_p/2}^{T_p/2} x(t + \tau) \cdot x(\tau)\, d\tau$$

Power $= R_{xx}(0)$

The power spectral density $S(f)$ is the Fourier transform of $R(t)$ and, because $R(t)$ is periodic, $S(f)$ is an ideally sampled function. The topic is pursued more fully in section 2.5.1.

It is also possible to define the cross-correlation between two functions. Its Fourier transform is the cross-spectrum, e.g.

$$\phi_{xy}(t) = \int_{-\infty}^{\infty} x(t+\tau) \cdot y(\tau)\, \mathrm{d}\tau$$

$$G_{xy}(f) = \mathscr{F}[\phi_{xy}(t)] = X(f) \cdot Y(f)\star$$

Similar results apply to periodic signals.

The definition of the autocorrelation function can be generalized to encompass complex time domain waveforms by conjugating the second function in the correlation integral.

Appendix 6

Useful Trigonometric Identities and Series Expansions

$$e^{\pm j\theta} = \cos\theta \pm j.\sin\theta$$

$$\cos\theta = (e^{j\theta} + e^{-j\theta})/2$$

$$\sin\theta = (e^{j\theta} - e^{-j\theta})/2j$$

$$\sin^2\theta + \cos^2\theta = 1$$

$$\cos2\theta = \cos^2\theta - \sin^2\theta \; = \; 2\cos^2\theta - 1 \; = \; 1 - 2\sin^2\theta$$

$$\sin2\theta = 2\sin\theta\cos\theta$$

$$2\sin A\sin B = \cos(A-B) - \cos(A+B)$$

$$2\cos A\cos B = \cos(A-B) + \cos(A+B)$$

$$2\sin A\cos B = \sin(A-B) + \sin(A+B)$$

$$2\cos A\sin B = -\sin(A-B) + \sin(A+B)$$

$$(1+x)^n = 1 + nx + \frac{n(n-1)}{1.2} x^2 + \frac{n(n-1)(n-2)}{1.2.3} x^3 \ldots \qquad |x| < 1$$

$$(1+x)^{-1} = 1 - x + x^2 - x^3 + \ldots \qquad |x| < 1$$

$$\ln(1+x) = x - \frac{x^2}{2} + \frac{x^3}{3} - \frac{x^4}{4} + \ldots \qquad |x| < 1$$

$$\sin x = x - \frac{x^3}{3!} + \frac{x^5}{5!} - \frac{x^7}{7!} + \ldots$$

$$\cos x = 1 - \frac{x^2}{2!} + \frac{x^4}{4!} - \frac{x^6}{6!} + \ldots$$

$$\tan x = x + \frac{x^3}{3} + \frac{2x^5}{15} + \frac{17x^7}{315} + \ldots \qquad |x| < \pi/2$$

$$\tan^{-1} x = x - \frac{x^3}{3} + \frac{x^5}{5} - \frac{x^7}{7} + \ldots \qquad |x| < 1$$

Index